Food Packaging Systems

The Author

Pramod K. Omre is a Professor at the Department Post-Harvest Process and Food Engineering, G.B.Pant University of Agriculture and Technology, Pantnagar, Uttarakhand, India. He obtained his M.Tech and PhD (Process & Food Engg) from G.B.Pant University of Agriculture and Technology, Pantnagar. He has a more than twenty years teaching, research and extension experience. He has published three book chapters and thirty research papers in different national and International journals. Participated at International and National conferences. He has designed and developed the Multi Fruit Grader, Jaggery Chocolate and Dhal Mill. Life member of various engineering society and journals, Advisor of engineering journals, Scientific panel member of FSSAI (Ministry of Health and Family Welfare).

Suman Singh is a Post-Doctoral Fellow at Department of Packaging, Yonsei University, South Korea. She obtained her B.Tech (Agri.Engg.) from G.B.Pant University of Agriculture and technology, Pantnagar in 2009, M.Tech (Process & Food Engg) with gold medal from Sam Higginbottom Institute of Agriculture, Technology & Science, Allahabad in 2011 and PhD (Process & Food Engg) from G.B.Pant University of Agriculture and technology, Pantnagar in 2014. She has published three book chapters and eleven research papers in different national and International journals and thirty conference proceeding.

Kirtiraj K. Gaikwad is PhD (Packaging) Scholar at Department of Packaging, Yonsei University, South Korea. He obtained his B.Tech (Food Science) from Dr Panjabrao Deshmukh Agriculture University, Akola, India in 2009, M.Tech (Food Tech) from Sam Higginbottom Institute of Agriculture, Technology & Science, Allahabad in 2011 and MS (Packaging) from Michigan State University, USA in 2013. He has published three book chapters and ten research papers in different national and International journals also twenty five conference proceedings.

Sandhya Madan Mohan is Head and Asstt Professor at the Department of Home Science Bhilai Mahila Mahavidyalaya, Bhilai Pt. R.S.S.University. Raipur (C.G.). She obtained her M.Sc.(Food and Nutrition) from Rani Durgavati University Jabalpur and Ph.D. from Pt. R.S.S.University Raipur (C.G.). She has more than 34 years of teaching experience. She has published 4 books, a story book for Neoliterates, 4 book chapters, published 9 research papers in International and National journals and 27 papers in seminar conference proceedings. She had been Chairperson-Board of Studies Home Science, Subject Expert-Board of Studies-Home Science Sarguja Uni. Ambikapur and Board of Studies, Home-Science, Digvijay College Rajnandgaon.(C.G.).

Food Packaging Systems

by

Pramod K. Omre

Suman Singh

Kirtiraj K. Gaikwad

Sandhya Madan Mohan

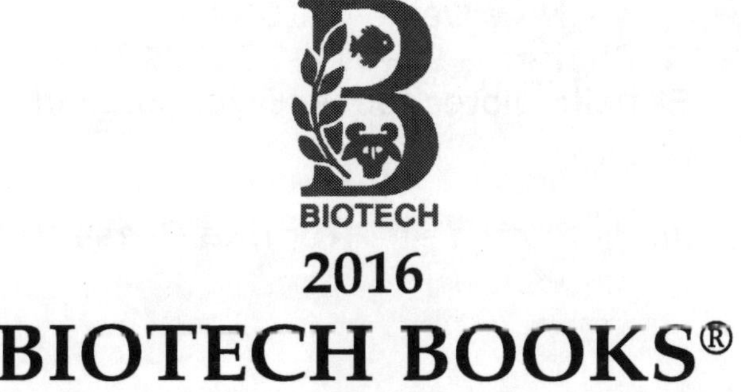

2016

BIOTECH BOOKS®

ISBN 978-81-7622-365-2

Published by: **BIOTECH BOOKS®**
4762-63/23, Ansari Road, Darya Ganj,
New Delhi - 110 002
Phone: +91-011-23262132
E-mail: biotechbooks@yahoo.co.in

Digitally Printed at : **Replika Press Pvt. Ltd.**

PRINTED IN INDIA

This book is dedicated to

OUR FAMILIES

Preface

Almost all food, food products and fuel commodities go through a number of post-harvest processing operations in that one of important operation is packaging. Packaging for consumer products is an area where supply and demand is continuously changing due to the development of an international food market and adaptation to consumer, distribution, legal and technological requirements. The world's total food production has more than doubled over the past fifty years due to improved methods in animal husbandry, the use of advanced seed varieties and crop protection products that boost crop yields and quality. Mass production of packaged food has been enabled by technological innovations in food production, processing and logistics with packaging playing a key role. The economies of scale involved and the intense industrial competition have made many products more affordable. Consumer demand for pre-packaged food continues to increase in advanced economies and a growing global population is also fuelling the demand. This is increasingly the case in newly industrialised countries experiencing rapid urbanisation.

This book informs the reader about major packaging material, food preservation techniques and product quality and shelf life, and the logistical packaging, packaging processes, necessary for a wide range of packaging presentations. It is essential that those involved in food packaging innovation have a thorough technical understanding of the requirements of a product for protection and preservation, together with a broad appreciation of the multi-dimensional role of packaging. This book sets out to assist in the attainment of these objectives by informing designers, technologists and others in the packaging chain about key food packaging technologies and processes. The authors are hopeful that this book will serves as good teaching aid in the field of food packaging technology.

Pramod K. Omre

Suman Singh

Kirtiraj K. Gaikwad

Sandhya Madan Mohan

Contents

1

Introduction

1.1 Packaging

Anything and everything around us today is packaged. Packaging has become a part of our lives. We are not able imagine a life.without packaging. What is packaging and do we really need it? This question can be answered only when we really imagine our life without packaging. The importance of anything is realized only when we are devoid of it. When we are in the shade we do not know the importance but when we go out in the hot sunshine then we realize the importance of shade. Same thing holds good for packaging.

During the days of Adam and Eve there was no need for packaging. Wants were limited and all those wants were available easily. But as civilization developed, mankind moved from place to place and mankind's wants also increased. Thus the concept of packaging took birth.

About a decade back, drinking water was available in plenty. Even we travel in trains we used to have the water from each station. Today we are not sure of the water in the station and hence drinking water is packaged. We all carry a bottle of "Bisleri" or "Aquafina". We want chilled water hence we have bottled water in freezer and have it. Fortunately, today the air we breathe is not packaged. But then for a person climbing the Himalayas needs packaged air (Oxygen). Why? It is not available in required quantity. Its distribution is not enough. So, packaging is *not an absolute need. But a derived need*. By this it is meant that, you have a derived want and that needs to be fulfilled hence it is a derived need. You need mangoes in the non-mango season, so it has to be packaged.

It is mankind's wants today that has resulted in packaging. You need a Sony Television from Japan; hence it has to be distributed from Japan. It cannot be just sent as-is because distribution will fail to deliver the goods in good and safe condition hence packaging is needed. *"When distribution fails packaging steps in"*.

1.2 Functions of Packaging

Packaging can be described as covering the product with one or more suitable materials for ease in handling, transportation and marketing. Packaging not only differentiates one brand from another but also, at times, gives a preview of the product being sold. In technical words, packaging is defined as 'the science, art and technology of enclosing or protecting products for distribution, storage, sale, and use.' Packaging also refers to the process of design, evaluation, and production of packages.

Most of the time, packaging is accompanied with attractive and informative labelling. A package label is any written, electronic or graphical message on the container of the packaged product. The role of packaging continues from the coordinated system of preparing goods to the end use.

Packaging contains, preserves, protects the product during its transport, and informs the customers about the properties of the product during its sales. With the passage of time, packaging industry and packaging techniques have undergone drastic changes. The stress has always been at reducing the after-use waste, reusing the containers wherever possible and recycling the waste to a maximum extent possible.

Packaging does a lot of functions, but the critical functions are as follows,

Contains - It holds the products in a specified volume of the bag or cartons, pack. Small objects are typically grouped together in one package for transport and handling efficiency. Alternatively, bulk commodities (such as salt) can be divided into packages that are a more suitable size for individual households.

Protects – Protection of the objects enclosed in the package from shock, vibration, compression, temperature, etc.

Preserves - A barrier from oxygen, water vapor, dust, etc.

Identifies – Identifies one product from another through its labeling. A bottle of water and bottle of spirit can be identified by a package.

Educates - Information on how to use, transport, recycle, or dispose of the package or product is often contained on the package or label.

Dispenses - Features which add convenience in ease of dispensing, dosing, distribution, opening, re-closing, use, and re-use.

1.3 Packaging Material

There a number of materials used for packaging in different forms. These materials are used either singly or in combination to achieve a particular function. The basic classification of the materials used is as follows:

Table 1.1: Different Packaging Materials and its forms

Material	*Forms*
Wood and Wood based	Cases and crates, Tea chests, Plywood chests
Paper and Paperboard	Bags, Envelopes, wrappers, cartons
Glass	Bottles, Jars, Containers
Metal (MS, TinPlate, Zinc Plate, Aluminium)	Drums, Tubes, Cans, Tubes, containers, Pails etc.
Plastics	Bottles, Caps, Drums, Pouches, Bags, Woven sacks etc.
Combinations	Laminated pouches, retort pouches, canisters etc.

1.4 Types of Packaging

a) Flexible - Soft and pliable package forms

b) Rigid - Hard and Rigid package forms

c) Semi Rigid - In between hard and flexible package forms

1.5 Categories of Packaging

Human needs to consume a product are plentiful and so are the packaging types. For example – there is transport package or distribution package which is the package form used to ship, store, and handle the product or inner packages. There is consumer package, which is directed towards a consumer or household. In relation of the product type being packed, there is medical device packaging, bulk chemical packaging, over-the-counter packaging, retail food packaging, military materiel packaging, pharmaceutical packaging etc.

Those who handle the product along the way need different labelling and packaging than the final user. For the ease of categorization, packaging is now categorized on the basis of layers, that is, primary, secondary and tertiary.

1.5.1 Primary Packaging (Unit Packaging)

Primary packaging contains the smallest quantity of a product for final sale or use. It is the package, which is in direct contact with the contents. The primary packaging that contains the product not only catches the customer's attention, but also creates a desire to buy the product and inspire the customer's confidence to buy the product repeatedly. The customer should feel a sense of satisfaction right from the feel of packaging. Primary packaging aims mainly at marketing purposes.

The point to be taken special care of while designing the primary package is to use descriptive titles for the product - not necessarily creative. Many people go into a retail store looking for a product, but do not necessarily have a specific product in mind. You need to communicate your function and benefits to them quickly and effectively. Graphics and slogans on the package should reflect the functionality of the product.

1.5.2 Secondary Packaging (Intermediate Packaging)

It is the packaging outside primary packaging – usually used to group primary packages together. For example, the family packs of the chocolates available in super markets, decorated carton or gift box are common examples. Secondary packaging, sometimes, is also called intermediate packaging. Since not all products use intermediate packaging, the following factors can be used to determine when intermediate packaging is needed:

1. How the product will be distributed,
2. How the product will be merchandised,
3. Whether the finished products will be sold in kits.

1.5.3 Tertiary Packaging (Bulk Packaging)

It is used for bulk handling, stockroom storage and transport shipping. It is the outer most level of packaging. Generally, other packages are shipped\transported with them, and they are designed to withstand normal transportation stresses. The corrugated, brown carton is the most familiar example of tertiary packaging.

1.6 Product Characteristics

One of the most important factors that one has to consider during the development of the package is the product characteristic. The characteristics of the product include their physical characteristics and their physio-chemical characteristics.

While we decide about the package all we need to know is the answer to the question: "How the product can be contained and how can the product get damaged or deteriorate." The things to be noted are as follow

1.6.1 Physical State of the Product

We should understand the general properties of the product such as whether it is solid, liquid or a gas. If the product liquid whether it is viscous or thin. If solid whether it is in the form of powder or it is a solid block. If gas the properties of the gas etc.

Size and Shape

To determine the dimension and the contour of the package one should know the dimensions of the product and the allowances to be given for inserting the product in the pack. For example if it is liquid which has a tendency to expand one has to give allowance for the normal expansions during storage and transportation and provide for head space.

Weight and Density

The weight and density of the product determine the type dimensions and cross-sectional dimensions of the packaging material to be used.

Centre of Gravity

For the pack to be stable during the storage and transportation, knowing the center of gravity is very important. The center of gravity should be as nearer to the center of the box as possible. Otherwise the pack would have a tendency to tilt and fall during storage and transportation.

Weaknesses

A product may have different components and parts. All these parts will be susceptible to damage in varying degree. Some parts for example such as a thin tubes or coils in machine may need additional protection from inside.

Strengths

Some of the products such as diesel engines or compressor or many other engineering goods have a base with bolting holes. These are normally the strongest part of the product and the pack is bolted to the product through this base. Its susceptibility to damage due to the climatic and environmental hazards.

Product Package Compatibility

Product package compatibility is important for all products, but is more relevant to food and pharmaceutical products. Product package compatibility essentially means that there should not be any interaction between the product and the package. For example some mineral oils and fats can react with polyethylene. Some papers and adhesive tapes if used on metallic parts may start reacting with the product. Some additives in the plastics may cause odors when used for packaging of food items.

Fragility of the Product

Fragile objects are more susceptible to damage than the sturdy products. It is therefore necessary to understand the damage causing shocks and vibrations during the distribution and one should pack the product in such a way so as to protect the product safely during its entire distribution cycle.

Corrosion

Corrosion is another factor which one has to consider during the packaging. Even if the pack is designed to protect the product from corrosive atmosphere outside, proper care should be taken to ensure proper product preparation before packing and ensuring that the atmosphere inside the package does not cause corrosion of the product.

Perishable Items

Here again the product preparation is essential. For example the microbe levels in the food items should be below a certain level. Also the causes for the accelerated decay of the product should be understood and the pack should be so designed. For example some product the deterioration would be accelerated due to sunlight. There for it is essential that these products should be packed with opaque packaging materials. Thus before we design a pack it is necessary to know all details and the characteristics of the products. Many a time a little relocation/modification of the product for example, would ease the packaging requirement to a great extent.

2
Packaging Material

In order to protect the product from the various hazards, the packaging materials should have the requisite properties. Also the packaging materials which run on fast automatic and semi-automatic machines should have sufficient strength and other properties to with stand the pressure from these machines.

The packaging materials vary from the basic paper and paper products to metals and plastics and combinations of these. Also there are varieties of forms and sizes. Each of these materials have different characteristics and properties. Therefore it is very important to understand the properties and the material characteristics of the various packaging media.

Paper

- Paper is a hygroscopic material – it absorbs moisture in high humid condition and it loses moisture in dry condition.
- The mechanical properties of paper is influenced by the moisture content of the paper
- Mechanical strength and elongation are generally low.
- Paper provides stiffness and shape to the pack.
- Paper offers an excellent printing surface.
- Many varieties of paper for special application – grease proof paper, glassine paper etc.
- Susceptible to tear and puncture.
- Eco friendly

Plastics

- Variety of materials – each varying in properties.
- High mechanical strength and high elongation.
- Plastics in general (with a few exceptions) are not moisture sensitive.
- It can be used both in the flexible forms and the rigid forms.
- Heat sensitive.
- Can be moulded into shapes through various processing techniques.
- It is impermeable to water ingress but not to moisture.
- The moisture or gas ingress depends on the type of plastics, the thickness of the materials, the atmospheric conditions etc.
- Different plastics can be combined to form a composite material to achieve the desired properties for the product protection.
- Surface is polar in nature and hence direct printing is difficult in most cases.
- Generally plastics have good tear, puncture and abrasion resistant properties.
- Generally they are translucent to clear but can be made opaque by addition of pigments.
- Recyclable under normal uncombined state.

Metals

- Very strong
- Impermeable material
- Good printing surface
- Susceptible to erosion unless lacquered or coated
- Susceptible to denting
- Can be used as thing foils to improve the permeation rates in plastic laminates
- Can be retorted / sterilized easily for food preservation.
- Particularly useful for food products which require long shelf life.
- Can withstand internal pressures
- Can be used for hot filling – Food products
- Manufacturing and filling can be done at high speeds.
- Eco friendly

Glass

- Chemically inert
- Non-permeable

- Transparent – can be colored
- Moldable to any shapes
- High strength
- Light weight
- Highly brittle
- Raw materials for manufacture are easily available
- Decoration through special process (frosting and ceramic printing)
- Eco friendly

Wood

- Hygroscopic
- Strong
- Properties depend on the moisture content in timber
- Abrasive
- Good hardness
- Inherent defects
- Susceptible to fungus and microbial attacks
- Dry wood susceptible to warping and bending
- Fabrication of wood into boxes and crates is manual
- Wood is becoming scarce.

Jute/Hessian

- One of the oldest form of sacking
- Strong
- Natural fiber
- Made from renewable source as compared to the synthetic fibers
- Being an agricultural product the availability depends on the climatic variations.

After application the sealant should be "aged" well so that the compound does not deteriorated during storage. Otherwise the air leakage will start and the product starts deteriorating.

3
Paper

3.1 History Note

- The paper was first made in China (BC 105)
- The Muslims around 950 brought the papermaking to Spain.
- 1st publication of books 1450
- Newspaper publication – Europe 1609
- 1st Paper Mill in Philadelphia 1690 (single sheet production)
- The Foundrinier brothers in England patented the continuous paper making development by Mr. Nicholas – Louis Robert.
- The cylinder type machine was invented by John Dickenson and installed near Philadelphia (1817)

3.2 Structure of Paper

The paper and board constitutes almost 60 per cent plus in the total tonnage of packaging materials. The wood and other natural cellulosic fiber containing materials are the basic sources of Paper. Depending upon the source, namely wood or agricultural waste (bagasse, straw, jute) or bamboo, the quality of the constituents mainly cellulose fiber content and characteristics vary. The final quality of paper depends upon the process, the additives and the coating and the purpose of the finished sheet.

Wood is made of about 50 per cent cellulose, 30 per cent Lignin and 20 per cent carbohydrates (xylan and mannan) along with resins, gums and tannins. The cross section of wood fiber when enlarged will show an empty space called "Lumen" followed by layers of fibrils known as Lamella. The lignin is in the middle lamella. The fibrils are made of micro fibrils. These in turn are composed of chains of cellulose molecules (about 3mn in each micro fibril) along with short chain semi cellulose molecules and other polymeric residues.

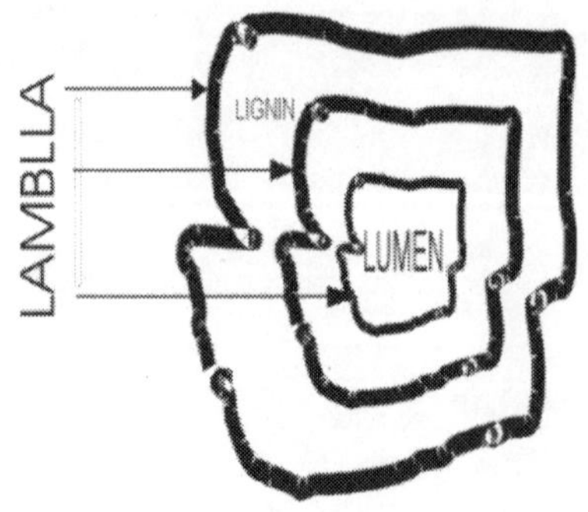

Softwood is the term used for the coniferous or needle bearing trees and has fiber length running up to 6.25mm. Hardwood is the deciduous trees whose fiber length is lesser than the softwood *i.e.* 2.5mm (max). In view of this, use of hardwood make finer, smoother sheet with less strength paper compared to softwood. Spiral winding called the cambium layer encloses the lamellae or layers of fibrils.

3.3 Basis in Pulp and Paper Making

- In the pulping process with water and chemicals – the fibrils swell, break the cambium, extend out and interlock with the hairy surface of other fibril.
- The water used gets in between lamellae and cause the swelling rather than getting into the cellulose molecules crystalline structure.
- Water molecules are held much tightly in the layer adjacent to the cellulose.
- In the pulping process, the lignin gets removed along with other soluble short chain polymers. Lignin present is highly hydrophilic in nature and also imparts color along with the tannins. Hence the maximum removal of lignin is aimed at during the pulping process.
- In the pulping process, the fibers are suspend in large amount of water, water gets removed. Fibers get deposited on the screen acting as a filter for the suspended particles.
- A more uniform sheet is tend to form with the principle thin areas have a higher rate of flow and hence fiber move in their direction.
- The course fiber being heavier settles more quickly than the finer ones. This phenomenon leads to the fact that the wire side is rougher than the top (felt).
- The forces which hold paper / fibers together and resist delamination are largely chemical and where the fibers cross, there is actually a hydrogen bounding.
- At the time of formation on the screen, the fibers have 60 per cent water within them. Hence water has to be driven out by heat rather than by squeezing.
- Denser sheets have considerable shrinkage.

3.4 Paper Making Process

3.4.1 Pulping

There are various ways of producing the pulp.

1. Ground wood pulp
2. Chemical pulp
3. Semichemical pulp

Groundwood Pulp

- Cheapest process
- Logs (after removal of back) are pressed against a grinder stone while water is sprayed over the stone. This enables filse to be carried. This process is used for softwoods only. All components in the wood are utilized in this process.
- Newsprint uses Ground wood or mechanical pulp.

Chemical Pulp

Wood (cooked) pulp + unwanted lignin and carbohydrates separation + chemicals.

a) Soda Process (Alkaline)

Wood + Sodium Hydroxide + SODA ASH

Generally used with hardwood pulps.

b) Sulphate Process – Kraft Process (Alkaline)

"KRAFT – Higher strength in German"

Wood + Sodium Hydroxide + sodium sulfide

PULP – BROWN IN COLOUR and DIFFICULT TO BLEACH

c) Sulphite Process

ACID REACTION

WOOD + $CaHSO_3$ or $MgSO_3$ + H_2SO_3

Used for softwood processing

Semichemical Pulp

Wood soaked in NaOH or Neutral sodium sulfide. (Soften the lignin and carbohydrate)

Process around in disk refiner

This process lower in cost uses primarily hardwood has high lignin retained.

Turns yellow when exposed to sunlight

Strength and stiffness are good (Hence used as corrugating medium)

Stage I: Preparatory Stage

Wood - Debarking - Cutting - Chipping

Stage II: Pulping (Digesters)

(Mechanical, Chemical and Semi-chemical processes) - Blow Tank

Stage III: Stock Preparation (Furnish)

(Beating and / or Refining) - Sizing - Use of fillers

Stage IV: Sheet Making

(Fourdrinier machine or Cylinder machine Combination of both)

Stage V: Finishing

(**PULPING in a nutshell** – Digesters are used for cooking at 170-180 degree Centigrade at 100 PS1 pressure (7Kg / Cm^2) with the wood chips for several hours and blown down into a 'blow pit' where it is washed, bleached). De-lignification of the pulp improves the strength and decreases the opacity.

Beater

Different designs of equipment are used to give this mechanical treatment. Hollander beater is one among them. The slurry is composed of 96 per cent water and 4 per cent solids. In the beater, the bundles of fibres are broken up, hydration of fibrils takes place and results in the increased surface area.

(Fibrillation is the process in which the cambium layer around each fiber is broken as the fibres swell and the fibrils open up). In the beater the slurry is passed through the clearance between the bars and plates, pass through a weir called backfall that helps in mixing of the stock.

Refining is combination of the same process in which equipment are of different designs (conical JORDAN refiners and disc refiners). Refining with new designs are replacing beaters.

Refining brings about the physical changes in the fibre structure. It gives rise to a potential binding or bonding between fibres in the formed web resulting in sheet strength.

In order to provide water resistance and ink holding properties materials such as Rosin, Starch, Papermaker's Alum are added in emulsion form in the beater.

Fillers improve texture, colour, opacity, stiffness, stability etc. (Kaolin, sodium silicate, casein, Titanium dioxide of particle size 0.4 to 5 micron). These are negatively charged. To help the retention of fine solids on the webduring drainage aids such as cationic starches, urea melamine resins are added.

The stock at this stage (pulp, sizing, fillers) is known as 'furnish', moves to the headbox getting ready for papermaking. The consistency of pulp at this stage is 1 per cent solid and 99 per cent water.

Both paper and board are made on similar machines.

Types of Machines for Paper Making

1. Fourdrinier

Mainly used for making thin paper below 0.3 mm. The dilute suspension of pulp is spread over a moving screen belt which allows water to pass through leaving a mat of paper. The moving belt is made of nylon or phosphor bronze and is stretched over a number of rolls of the system called by various names.

The web is transferred to the press roll where water content comes down from 80 to 60 per cent. It is then passed in and out of steam heated stream of rollers as many as 150 at times in several stocks, till the moisture content drops to 4 to 8 per cent. Now a days large size rolls are replacing a too many smaller size rollers.

Surface treatments are applied at times while paper is still on machine (like dipping the paper in starch solution and squeezing between rolls to improve printing and stiffness. The paper is passed over stack of chilled rollers to give compactness and smoothness. However excessive calendaring reduces the stiffness.

2. Cylinder Machines

This is used for making heavy grades of paperboard from waste materials with layer of high grade material outside.

It consists of a cylinder wire screen wound over a drum. This when revolves inside a vat containing pulp slurry suspension picks up layers of pulp, forms the web on a felt blanket. Several such layers are built up on the felt to make the thickness and this moves to the end. The high quality pulp on the top is "top liner" and opposite is back liner. The plies in between are called fillers.

3. Inverform

Combination of endless wires of FOURDRINER with multiple head boxes of the cylinder machine.

Paperboard

Paperboard is a thick paper based material. While there is no rigid differentiation between paper and paperboard, paperboard is generally thicker (usually over 0.25 mm / 0.010 in or 10 points) than paper. According to ISO standards, paperboard is a paper with a basis weight (grammage) above 224 g / m^2.

Paperboard can be single or multi-ply. Paperboard can be easily cut and formed, is lightweight, and because it's strong, it's used in packaging. Another end-use would be graphic printing, such as book and magazine covers or postcards. Sometimes it is referred to as *cardboard,* which is a generic, lay term used to refer to any heavy paper pulp based board.

In 1817, the first paperboard carton was produced in England. Folding cartons first emerged around the 1860s and were shipped flat to save space, ready to be set up by customers when they were required. 1879 saw the development of mechanical die cutting and creasing of blanks. In 1911 the first kraft sulphate mill was built in

Florida and in 1915 the gable top milk carton was patented and in 1935 the first dairy plant was observed using them. Ovenable paperboard was introduced in 1974.

Best Designs are a Result of

- Personal creativity plus
  - Knowledge and understanding of packaging materials, including:
- Structural properties
- Graphic capabilities
- Converting process
- Customer packaging systems
- Marketing objectives
- Distribution requirements
- Retail outlet expectations
- Needs and desires of end user
- How end user will use the product
- Many people may contribute to the design

Overall, the Design Must Provide

- Containment of product
- Protection of product
- Unitization for ease in handling through distribution
- Prevention of product spoilage
- Tamper evidence
- Consumer convenience
- Brand identification
- Communications for:
  - Instructions for product use
  - Coding for quality assurance, expiration dates
  - Dietary and nutritional information

Before Beginning Consider Three Important Areas

1. Converting or package manufacturing issues
2. Customer issues for filling and sealing
3. Consumer issues for convenience and performance

Converting Issues

- Design will easily move through converting plant
- Artwork can be reproduced to customer's satisfaction
- Thickness of paperboard with compensation for creases and tucks

- Waste and spoilage are minimal
- Package's life cycle is understood and will meet customer's expectations.

Grain Direction of Paperboard is Important

- Printing
- Automated gluing
- Reduce bulge
- Reduce shrinkage

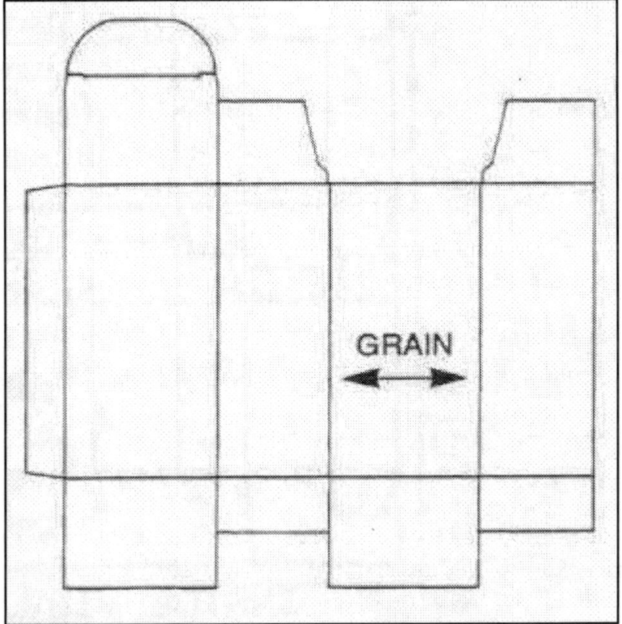

Several Methods to Determine Grain Direction

- Rely on pallet data sheet
- Cut a sample sheet, wet the sheet and allow to dry. As the sheet dries it will curl with the grain of the sheet in the long direction of the curl.
- Take two sample sheets, tear one in half from top to bottom. Take a second sheet and tear it from side to side. The tear that is the straightest will be in the grain direction or MD. The other tear will be ragged and may run in any direction.
- Tear two samples, one top to bottom, the other left to right. The direction with the least resistance is the grain direction

Corrugated Fiberboard

What is a Corrugated Fiberboard?

A corrugated fiberboard is basically made by bonding two linerboards and a corrugating medium together by an adhesive as shown in figure below. Therefore, the physical properties of a corrugated fiberboard depend upon the type of the flutes adopted, and the kinds of linerboard and medium used.

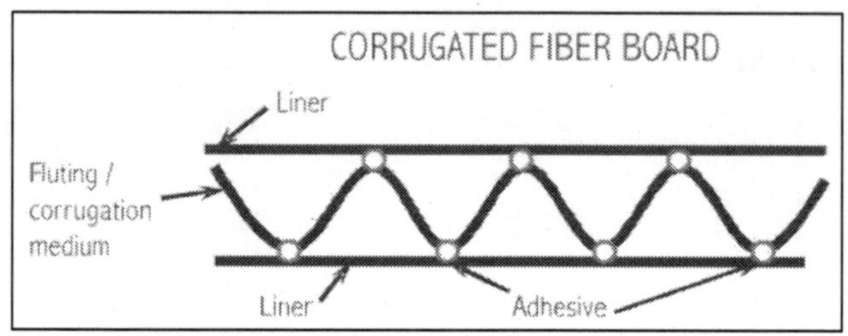

Base Paper of Corrugated Fiberboard

Paper used for making corrugated fiberboard is generally called base paper of corrugated fiberboard. It is classified as follows:

Adhesive for Corrugated Fiberboard

However excellent the quality of base paper of corrugated fiberboard may be, corrugated fiberboard with stable quality cannot be made quickly unless there is a good adhesive bonding the linerboard with the medium. Recently the production techniques for corrugated fiberboard have made remarkable progress, and machines have become wide and fast as a worldwide trend. Machines of 2,000 mm or more in width and 200 m/min in speed are now usually adopted. A key point for meeting such productivity is supposed to be in the adhesive, and at present, researches of the adhesive are made throughout the world.

The adhesive now used for corrugated fiberboard is based on corn starch. Fortunately, the starch which well suits the bonding mechanism of corrugator, *viz.* corrugated fiberboard making machines is low-priced and satisfactory enough also in performance. It is actually used throughout the world. Needless to say, the reason is that the chemical technology for perfect use of starch as an adhesive had been established.

The detailed mechanism is not explained here since it is too special. Anyway it is supported based on the following:

Adhesive
- → Starch (base material of adhesive)
- → Caustic soda (promoting the geletinization)
- → Water (adjusting the concentration and the viscosity)
- → Borax (increasing machinability and intensifying the starch film)

What is most seriously considered here is bondability, needless to say, and the next very important factor is the adaptability to the high speed machine, in terms of viscosity and concentration. Therefore, recently, the basic reforming of starch is studied, to make starch with high concentration and low viscosity. The quantity of corn starch used in our country as an adhesive for corrugated fiberboard is estimated to be about tons, judging from the production of corrugated fiberboard of last year. Considering that it is mostly imported and that it is a kind of foods, it involves various complicated problems.

Present Situation of Corrugated Fiberboard

The base paper used for corrugated fiberboard is purchased by weight, and the corrugated fiberboard produced are sold by area. Therefore the quantity of produced corrugated fiberboard is indicated by area.

Kinds of Corrugated Fiberboard

Corrugated fiberboard with various physical properties can be made by combining the types of flutes.

Type of Flutes Item	*Number of Flutes 30cm*	*Flute Height (mm)*	*Fluting Rate*
A-flute	34#2	(approx.) 4.6	(approx.) 1.6
B-flute	50#2	(approx.) 2.5	(approx.) 1.4
C-flute	40#2	(approx.) 3.5	(approx.) 1.5
E-flute	93#2	(approx.) 1.0	(approx.) 1.1

There are four types of flutes used for producing corrugated fiberboard, being classified as shown in in reference to the number of flutes per unit length (usually 30 cm) and flute height.

At present, the types of flutes and numbers of flutes are specified in JIS Z 1516 (corrugated fiberboard), but the flute heights are not specified.

These flutes are used to make corrugated fiberboard, and the following four kinds of corrugated fiberboard can be made, depending on how to use and how to combine the flutes.

Single-Faced Corrugated Fiberboard

A single faced corrugated fiberboard is made by bonding a corrugating medium on to a linerboard, and can be said to be a basis of corrugated fiberboard. However,it is little used directly to make boxes, but rather used mainly as paddings and cushioning materials.

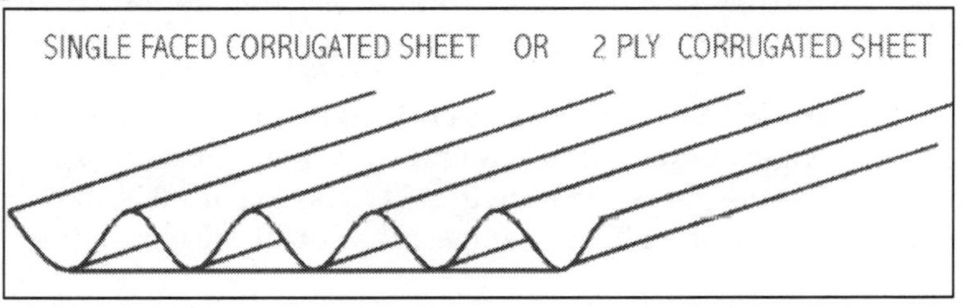

Double-Faced Corrugated Fiberboard or Single-wall Corrugated Fiberboard

A double-faced corrugated fiberboard is made by bonding a linerboard on to the flute tips of the medium of a single-faced corrugated fiberboard as shown in Fig below.

As indicated above, this corrugated fiberboard is called in two ways, depending on whether it is called in reference to the linerboard used or the corrugating medium. A double-faced corrugated fiberboard has different physical properties, depending on the type of flutes used. Therefore, the types of flutes are selected according to the natures of contents to be packages. In general, for hard and strong contents, B-flutes are used, and for soft and breakable contents, A-flute are used.

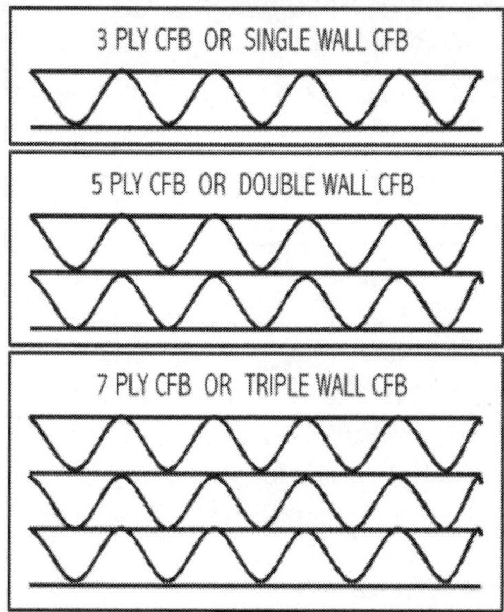

Double-wall Corrugated Fiberboard

A double-wall corrugated fiberboard has a structure with a double-faced corrugated fiberboard combined with a single-faced corrugated fiberboard. Usually, two types of flutes are used. Therefore, it is more structurally complicated, more intensified in physical properties and better in cushioning capability than the double-faced corrugated fiberboard. Thus, it is often used for packaging breakable contents and those which must be stored for a long time.

Triple-wall Corrugated Fiberboard

A triple-wall corrugated fiberboard has a structure with a double-wall corrugated fiberboard combined with a single-faced corrugated fiberboard. Three types of flutes can be used in combination, but one type of flutes may be used doubly. Therefore, it is structurally very complicated and very intensified in physical properties, with little directional since lengthwise and crosswise strengths are almost the same. At present, in our country, this is very little used, but in USA, it is often used in combination with pallet and skid, for packaging heavy cargoes, as a substitute for wooden box.

Top and Back Side of Corrugated Fiberboard

When a corrugated fiberboard is made in to a box, a beautiful side is brought to the top. The reason is that since a corrugated fiberboard box is normally printed on the surface, the surface must be, needless to say, as smooth and printable as possible to give a larger printing effect. How to distinguish the top of a corrugated fiberboard from the back side will explained below. The production process of corrugated fiberboard includes a single facer and a double facer. Both the production steps are greatly different in function, and therefore the top and backside of a corrugated fiberboard can be naturally classified.

Single Facer

A sectional view of the mechanism for producing a corrugated fiberboard by a single facer is shown in figure. A medium passing through a pair of upper and lower corrugating rolls heated to about 180°C gets the water evaporated by the heat and pressure applied at that time and the resin content in the medium is hardened and formed like waves. Immediately after that, the flute tips of the medium are coated with a certain amount of adhesive and are brought into contact with the back side of a linerboard fed in from another source, and both are momentarily heated and pressed between the flute tips of the lower corrugating roll and the pressed roll, to be glued with the adhesive put between the medium and the linerboard.

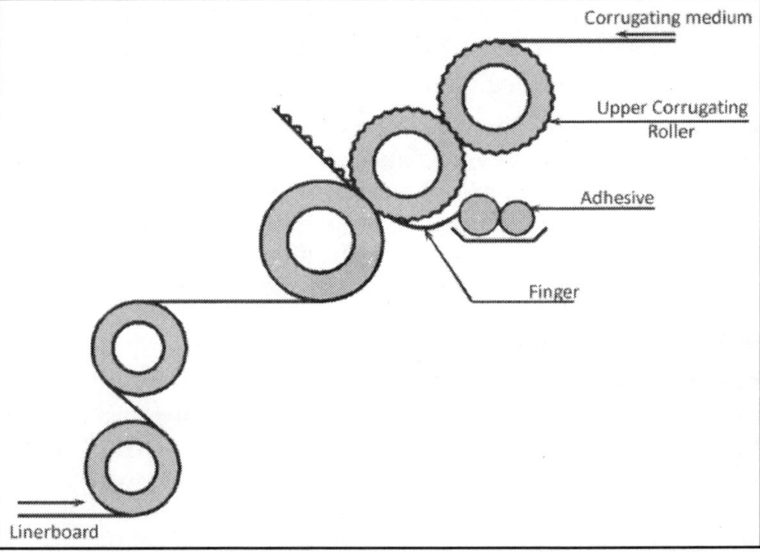

Therefore, the linerboard has stripes formed on the surface along the flute tips of the medium by the pressure applied at that time. The stripes are generally called a pressmark. The stripes uglily form some unevenness on the surface of the formed single-faced corrugated fiberboard. Thus, this side is decided as the backside of a corrugated fiberboard.

This phenomenon tends to appear more remarkably if the linerboard is thinner, as a matter of course.

Double Facer

A sectional view of the mechanism for producing a corrugated fiberboard by a double facer. In this gluing mechanism, the single-faced corrugated fiberboard made by the single facer should not be impaired as far as possible.

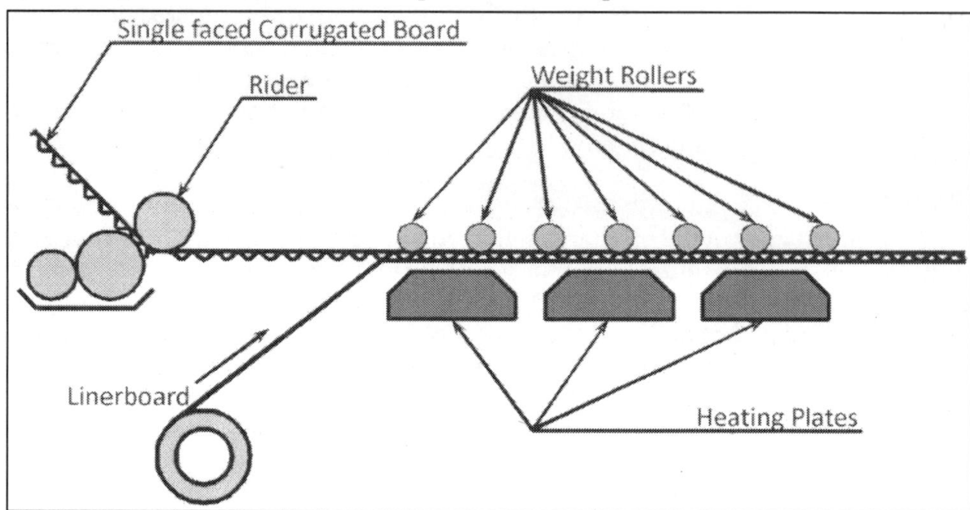

Therefore, at the time of applying the adhesive, the heating for gelatinization and pressure are very important. Especially, the pressure must be controlled to the minimum extent required for gluing.

The reason is that excessive pressure makes the flutes of the corrugated fiberboard defective, crushing the flutes or causing leaning, to seriously affect the quality of the completed corrugated fiberboard. However, in general, since the gluing mechanism is fundamentally different from that of the single facer, the surface of the corrugated fiberboard on the double facer side is flat and is finished to be very printable. Therefore, this side is decided as the top of the corrugated fiberboard.

Basic Quality of Corrugated Fiberboard

The essential features of corrugated fiberboard must be fully understood, to effect good corrugated fiberboard packaging. What on earth are the basic physical properties of corrugated fiberboard? The physical properties specified in JIS Z 1516 (Corrugated Fiberboard) are bursting strength and moisture content, and in addition as reference values, edge crush strength values are shown.

Moisture

With regard to moisture, since paper has hydrophilic groups, it has a nature of being liable to be chemically familiar with water, and absorbs and discharges moisture according to the change of the surrounding. Especially, corrugated fiberboards are surmised to be very sensitive in this regard. The moisture content of corrugated fiberboard cannot be said to be a physical property in a strict sense, but the behaviour of moisture in corrugated fiberboard must be well understood, since otherwise

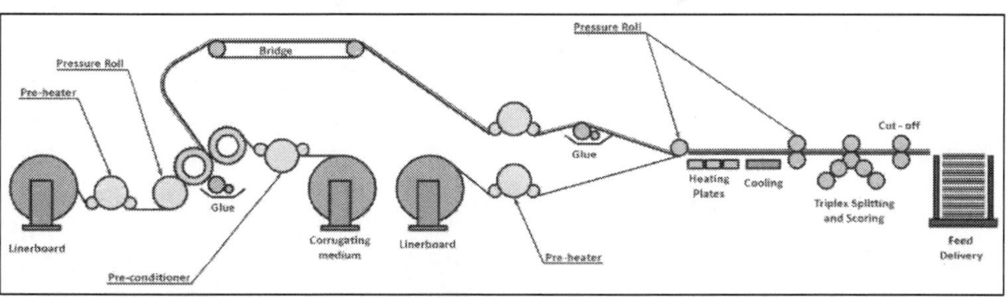

unexpected failures may occur. Since corrugated fiberboard are made of paper, the behaviour of moisture of corrugated fiberboard can be considered to be almost the same as that of paper. In general, if corrugated fiberboard are allowed to stand in the standard condition, *viz.* at 20°c and 65 per cent R.H. for more than 10 hours, they finally contain about 9 per cent of moisture. However, when they are actually used as corrugated fiberboard box, the relative humidity changes always, and especially in the rainy season, it may rise near to 100 per cent. In this case, corrugated fiberboard box absorb much moisture in air, and the moisture content may increase to about 18 per cent. That is, it may reach about twice that in the standard condition, and in this case, the physical properties of the corrugated fiberboard box drop considerably. The moisture content of corrugated fiberboard is measured according to JIS P 8127 (Methods of Measuring the Moisture Content of Paper). As for the size of samples, it is specified to measure with a size of 250 x 400 mm in JIS Z 1516 (Methods of Measuring the Moisture Content of Corrugated Fiberboard).

4

Metal

Introduction

The metals generally used in packaging includes, Aluminium, Tinplate, Galvanized sheets and Mild steel. Broadly this can be classified as:

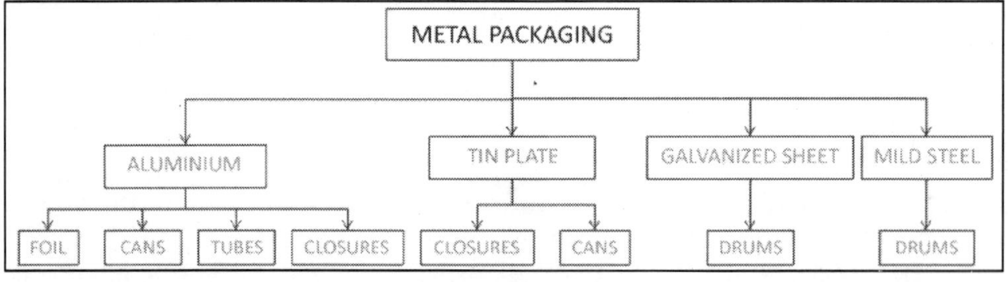

What is Aluminium?

It has low density compared to other metals: about one-third the density of iron and steel, less than one-third that of copper, and one-quarter that of lead and silver. Aluminium is the earth's third most abundant element (after oxygen and silicon) and the most abundant metal in the earth's crust (8 per cent by mass). Aluminium bearing compounds have been used by man from the earliest times.

Aluminium

This is used in the form of

- Aluminium Foil

- Aluminium Cans
- Aluminium Collapsible tubes

For all the above the purity of aluminium should be minimum 99.97 per cent or else the packaging cannot e formed. Pottery was made from clays rich in hydrated silicate of aluminium and at one point in history aluminium was so valuable that emperors and the wealthy preferred cutlery made by aluminium instead of gold. (Napolean's dinner set)

Today more aluminium is produced each year than all other nonferrous metals combined.

Production of aluminium has two different routes:

- From ore
- From recycling

Aluminium is easily formed: it can be drawn into thin wire and rolled into sheet to make airliners, or coils to make drinks cans, or foil. It can also be cast into engine components, and extruded into complex shapes such as window frames.

The raw material for Aluminium is BAUXITE. Bauxite is a reddish rock, containing aluminium oxide, together with oxides of silicon, iron and other metals. It was discovered by Pierre Bertier in 1822, near a village in southern France called Les Baux. It is mined, using an opencast strip method, mainly

in Australia, West Africa and the West Indies. Only a small proportion of the mining is in tropical rain forests. Out of total 14,000,000 square km of rain forest in the world, mining is in an area of just 5 square km at any given time.

How is the Metal Made?

Electrolytic process still used today was discovered in 1886 by Charles Martin Hall in America and Paul Heroult in France. Small-scale processes had enabled aluminium to be produced, for mainly decorative pieces, in time for the Paris Exposition in 1855.

Production process involves two main stages:

1. Making of Alumina powder (Al_2O_3)
2. Electrolytic smelting of Alumina (Al_2O_3)

Aluminium Production Stage 1

In the first stage crushed bauxite is mixed with hot caustic soda, which dissolves the aluminium oxide. Impurities can be filtered out and the caustic solution is cooled to crystallise the dissolved aluminium oxide into a white sand-like powder. This is called refining of Alumina, or aluminium oxide (Al_2O_3) in a caustic soda/high temperature process. The resulting fine white granular substance has a similar appearance to flour.

Aluminium Production Stage 2

Second stage is the Electrolytic smelting process. This involves mixing the alumina with a molten cryolite (sodium-aluminium fluoride) in a pot lined with pitch and coke. Powerful electric current is passed through liquid that splits aluminium oxide into aluminium and oxygen. The lining acts as a cathode so that when a carbon anode is inserted into the molten bath, and a strong electrical current passed through the pot, the alumina breaks down into molten aluminium and oxygen. The molten aluminium is syphoned off into furnaces where other elements are added to produce particular alloys, before being poured into moulds to form ingots.

It takes four tonnes of bauxite to make two tonnes of alumina, which makes one tonne of aluminium, consuming around 14,000 kwh electricity.

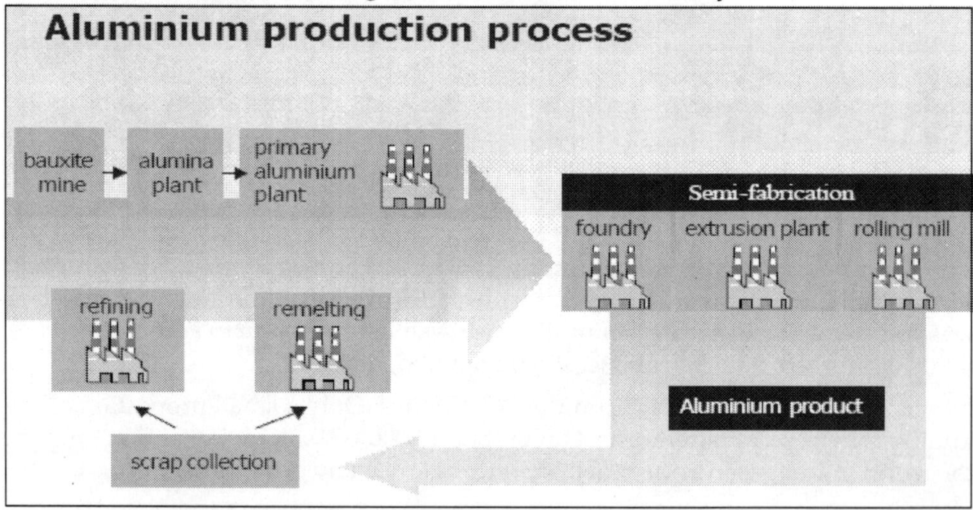

Aluminium is delivered to the manufacturing industry in three main categories of shapes:

- Flat rolled products, *i.e.* sheet and foil
- Long products: extrusions, bars, rods and wires
- Castings

In most cases by far aluminium is alloyed with addition of other metals to make an array of luminium alloys with different properties. The main alloying ingredients are manganese, magnesium, silicon, zinc and copper. The mechanical properties (yield strength and elongation) cover a wide range, in addition to the common undamental properties (lightness, corrosion resistance, electrical and thermal conductivity, brightness, etc).

It is an excellent conductor of heat and electricity, and has a relatively low melting point (660 °C).

And it is corrosion-resistant, due to a thin, but tough, film of oxide that forms on its surface when exposed to the atmosphere. And it is because aluminium can be recycled time and again without loss of properties, and because the energy savings therefore accumulate, that recycling is so important. Around 73 per cent of aluminium used in the world has been previously recycled.

Aluminium Collapsible Tubes

A collapsible tube is defined as a cylinder of pliable metal that can be sealed in such a manner that its contents, although readily discharged in any desired quantity, are protected from contact with air or moisture. Products so packaged must flow under pressure low enough not to damage the tube,

but must be sufficiently viscous not to spill out of the tube. Commercial production of aluminum collapsible tubes by impact extrusion began in Switzerland in 1914 and in the United States in 1921.

Collapsible tubes are made by impact extrusion using materials which are DUCTILE. These tubes are SEAMLESS. Materials used are TIN, LEAD and ALUMINIUM. Tin and Lead were the materials used till 1960. Tinand Lead are not used now because of high cost. In Lead chemical inertness is an asset. Today, most impact extrusion is Aluminium.

Collapsible Tube Production Stages

- Impact extrusion
- Annealing
- Coating
- Decoration
- Packaging

Impact Extrusion

For impact extrusion METAL SLUG is used. PURITY 99.97 per cent. The inner diameter determines the orifice.

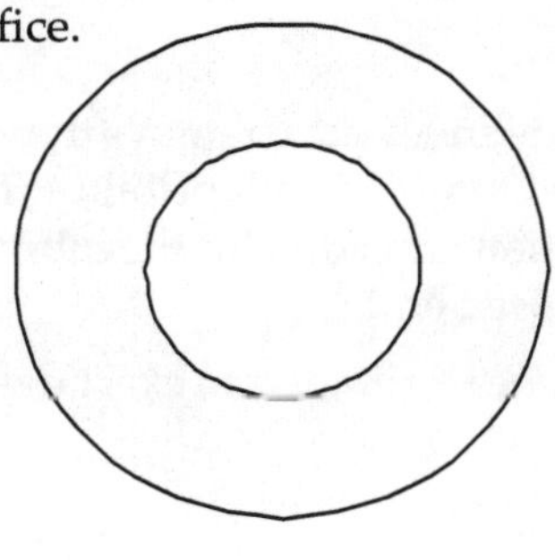

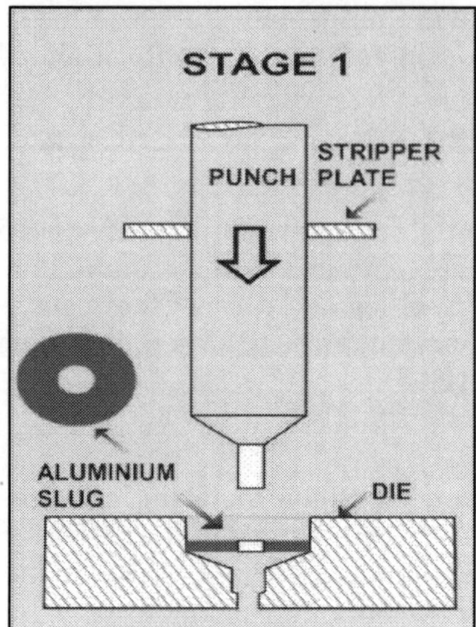

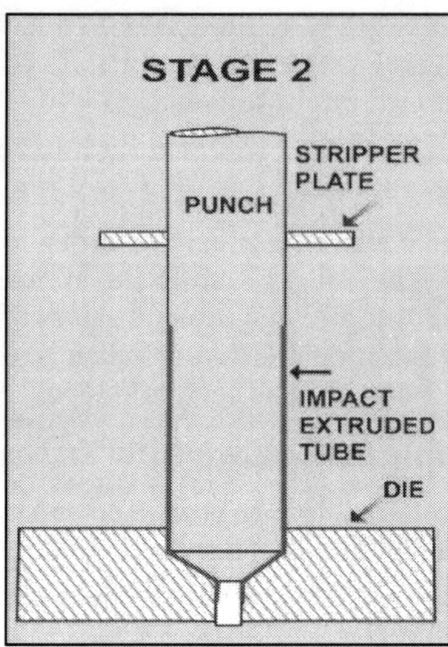

Stage 1

Metal slug is located on a shaped striking surface or anvil. A punch strikes slug with great force.

Zinc Stearate (powder) is used as lubricants to slug to facilitate extrusion and reduce heat.

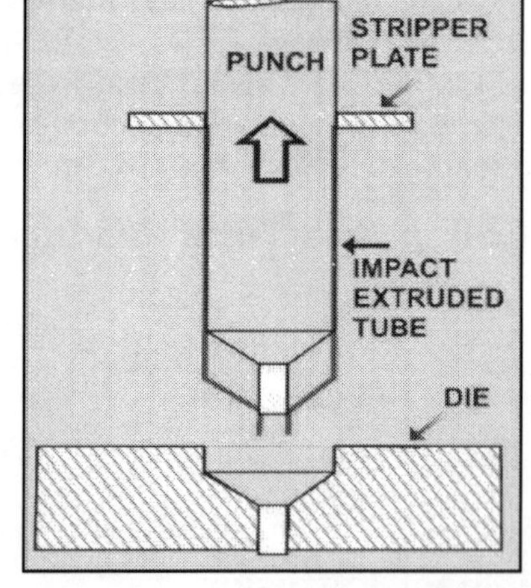

Stage 2

Under enormous impact

Pressure metal flows like a LIQUID straight upalong the outside of the striking punch forming a round cylindrical shape.

Stage 3

The tube height is governed by the thickness of the initial slug. Shoulders and tip are formed as a part of the process. Tubes with a thin web in the orifice will start with a solid slug. Force of impact work hardens and makes it stiff.

Annealing

After extrusion, the tubes are annealed to remove the work hardening and provide the softness or impness needed for good collapsibility. The degree of hardness that remains in the annealed condition is needed for the tube to maintain its shape and to hold the crimped fold at the closed end.

The non capping end of tube is trimmed to exact desired length.

Coating

Coatings are required inside some aluminum collapsible tubes to prevent corrosion by certain products. Even a superficial amount of corrosion, which might be tolerated for other applications, is objectionable in the tubes, because gas produced by the corrosion reaction causes the tube to swell. Consequently, coatings that provide a high degree of protection against corrosion are required. Even the outer surface needs a white surface coating.

Decoration

Decoration is needed for collapsible tubes to identify products packed in them. Tube outside needs a white enamel base coat and cured before decoation. The decoration is achieved by use of an artwork.This is printed by offset flexography using stereos. The decoration is cured in a oven. They are then capped.

Packaging

Collapsible tube is susceptible to denting because of its softness. These needs to be packed to avoid hitting against each other as well as from mechanical impacts during transportation. Generally they are packed in corrugated boxes with honey comb partitions. Boxes and partitions are often reused to reduce cost.

Parts of Tube

Dimensions of a collapsible tube is always on an unfilled tube.

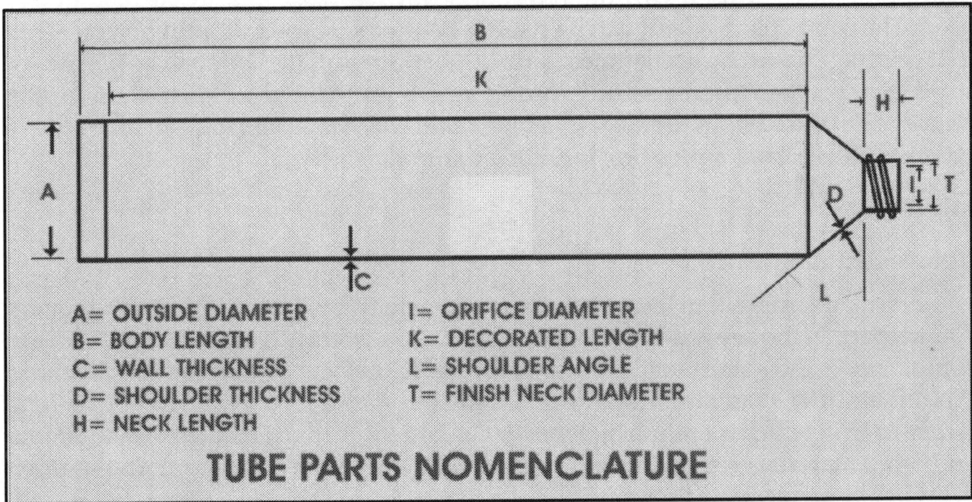

TUBE PARTS NOMENCLATURE

Types of Tube

There are a variety of aluminium collapsible tubes available depending upon applications. These variations are basically on the

1. Diameter
2. Length

3. Neck finish
4. Thread finish
5. Tips

Aluminium Collapsible Tube Availability

Most commonly used Tubes (diameters in mm). The length of the tube can be maximum **180mm**

10.00, 12.70, 13.50, 14.20, 16.00, **19.00,** 20.60, 22.00, 22.20, 25.00, **25.40,** 27.00 28.00, 28.60, **30.00, 31.70, 35.00,** 38.00, 40.00

Quality Control

1. Dimensions - Diameter, Length, Orifice, Thread, Shoulder height
2. Hardness Test
3. Decoration - Recoating, Adhesion

Aluminium Foil

Aluminum foil is aluminum prepared in thin metal leaves, with a thickness less than 0.2 millimetres (0.0079 in), thinner gauges down to 0.006 mm are also commonly used. The foil is pliable, and can be readily bent or wrapped around objects. Thin foils are fragile and are sometimes laminated to other materials such as plastics or paper to make them more useful. Aluminium foil supplanted tin foil in the mid 20th century. Metallised films ***are sometimes mistaken for aluminium foil, but are actually polymer films*** coated ***with a thin layer of aluminium.***

History

Before Aluminium Foil

Foil made from a thin leaf of tin was commercially available before its aluminium counterpart. In the late nineteenth century and early twentieth century, tin foil was in common use, and some people continue to refer to the new product by the old name. Tin foil is stiffer than aluminium foil. It tends to give a slight tin taste to the food wrapped in it, which is one major reason it has largely been supplanted by aluminium and other materials for wrapping food. The first audio recordings on phonograph cylinders were made on tin foil.

First Aluminium Foil

Tin was first replaced by aluminium in 1910, when the first aluminium foil rolling plant, "Dr. Lauber, Neher and Cie. and Emmishofen" was opened in Kreuzlingen, Switzerland. The plant, owned by J.G. Neher and Sons, the aluminium manufacturers, started in 1886 in Schaffhausen, Switzerland, at the foot of the Rhine Falls - capturing the falls' energy to produce aluminium. Neher's sons together with

Dr. Lauber discovered the endless rolling process and the use of aluminium foil as a protective barrier in December 1907. The first use of foil in the United States was in 1913 for wrapping Life Savers, candy bars, and gum. Processes evolved over time to include the use of print, colour, lacquer, laminate and the embossing of the aluminium.

Manufacture

Aluminium foil is produced by rolling sheet ingots cast from molten aluminium, then re-rolling on sheet and foil rolling mills to the desired thickness, or by continuously casting and cold rolling. To maintain a constant thickness in aluminium foil production, beta radiation is passed through the foil to a sensor on the other side. If the intensity becomes too high, then the rollers adjust, increasing the thickness. If the intensities become too low and the foil has become too thick, the rollers apply more pressure, causing the foil to be made thinner.

The continuous casting method is much less energy intensive and has become the preferred process. For thicknesses below 0.025 mm (0.00098 in), two layers are usually put together for the final pass and afterwards separated which produces foil with one bright side and one matte side.

The two sides in contact with each other are matte and the exterior sides become bright, this is done to reduce tearing, increase production rates, control thickness, and get around the need for a smaller diameter roller.

Some lubrication is needed during the rolling stages; otherwise the foil surface can become marked with a herringbone pattern. These lubricants are sprayed on the foil surface before passing through the mill rolls. Kerosene based lubricants are commonly used, although oils approved for food contact must be used for foil intended for food packaging.

Aluminium becomes work hardened during the cold rolling process and is annealed for most purposes. The rolls of foil are heated until the degree of softness is reached, which may be up to 340 °C (644 °F) for 12 hours. During this heating, the lubricating oils are burned off leaving a dry surface. Lubricant oils may not be completely burnt off for hard temper rolls, which can make subsequent coating or printing more difficult.

Properties

Aluminium foils thicker than 0.025 mm (0.00098 in) are impermeable to oxygen and water. Foils thinner than this become slightly permeable due to minute pinholes caused by the production process. Aluminium foil has a shiny side and a matte side. The shiny side is produced when the aluminium is rolled during the final pass. It is difficult to produce rollers with a gap fine enough to cope with the foil gauge, therefore, for the final pass, two sheets are rolled at the same time, doubling the thickness of the gauge at entry to the rollers. When the sheets are later separated, the inside surface is dull, and the outside surface is shiny. This difference in the finish has led to the perception that favouring a side has an effect when cooking. While many believe that the different properties keep heat out when wrapped with the matte finish facing out, and keep heat in with the matte finish facing inwards, the actual difference is imperceptible without instrumentation. The reflectivity of bright aluminium foil is 88 per cent while dull embossed foil is about 80 per cent.

Packaging Uses

As aluminium foil acts as a complete barrier to light and oxygen (which cause fats to oxidise or become rancid), odours and flavours, moisture, and bacteria, it is used extensively in food and pharmaceutical packaging. Aluminium foil is used to make long life packs (aseptic packaging) fordrinks and dairy products which enables storage without refrigeration. Aluminium foil laminates are also used to package many other oxygen or moisture sensitive foods, and tobacco, in the form of pouches, sachets and tubes, and as tamper evident closures. Aluminium foil containers and trays are used to bake pies and to pack takeaway meals, ready snacks and long life pet foods. Aluminium foil is widely sold into the consumer market, often in rolls of 500 mm (20 in) width and several metres in length.

It is used for wrapping food in order to preserve it, for example when storing leftover food in a refrigerator (where it serves the additional purpose of preventing odour exchange), when taking sandwiches on a journey, or when selling some kinds of take-away or fast food.

Aluminium Cans

Manufacturing of Two Piece Cans

1. Can Body

The two-piece beverage can is produced in a continuous process, which converts aluminium into cans. The process incorporates metal-forming, cleaning, internal and external treating and coating, and the application of final decoration.

Step 01: Aluminium coils are received at our plants in preparation for the manufacturing process.

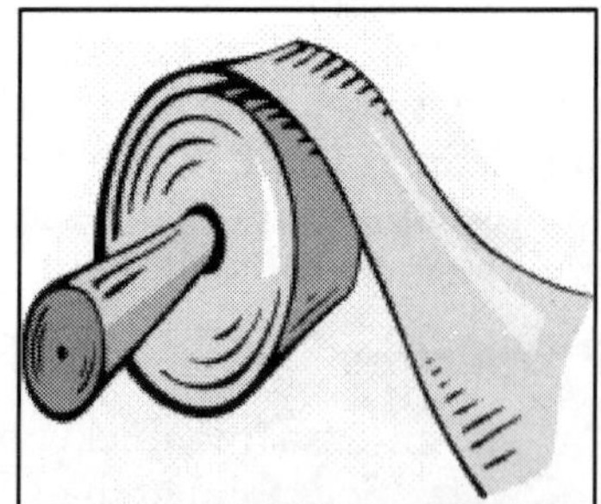

Step 02: The aluminium coil is cut into strips. The strip is lubricated with a thin film of oil and then fed continuously through a cupping press which blanks and draws thousands of shallow cups every minute.

Step 03: Each cup is punched through a series of tungsten carbide rings. This is the drawing and ironing process which redraws and literally thins and lengthens the walls of the cans into its final can shape.

Step 04: Trimmers remove the surplus irregular edge and cut each can to a specific height. The surplus material is collected and recycled.

Step 05: The trimmed can bodies are passed through highly efficient washers and then dried. This removes all traces of oil in preparation of the internal and external coating.

Step 06: The clean cans may be base coated externally with a clear or pigment base coat which forms a good surface for the printing inks.

Step 07: The cans pass through a hot air oven to dry the coating.

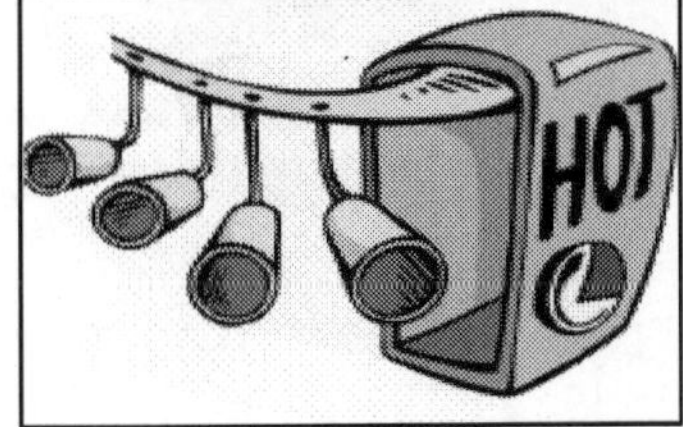

Step 08: The next step is a highly sophisticated printer/decorator which applies the printed design - up to six colours, plus a varnish.

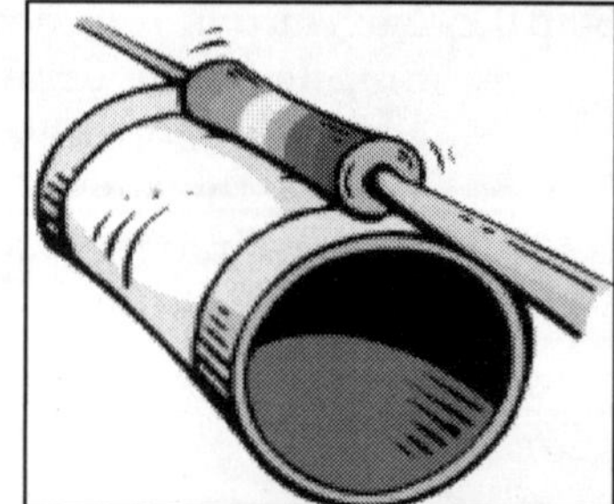

Step 09: A coat of varnish is also applied to the base of each can by a rim-coater.

Step 10: The cans pass through a second oven which dries the inks and varnish.

Step 11: The inside of each can is sprayed with lacquer. This special layer is to protect the can from corrosion and its contents from any possibility of interaction with the metal.

Step 12: Internal and external lacquered surfaces has already been dried in an oven.

Step 13: The cans are passed through a necker/flanger. Here the diameter of the wall is reduced or 'necked-in'. The top of the can is flanged outwards to accept the end once the can has been filled.

Step 14: Every can is tested at each stage of manufacture. At the final stage it passes through a light tester which automatically rejects any cans with pinholes or fractures.

2. Can End

Step 01: Can end manufacture begins with a coil of special alloy aluminium sheet.

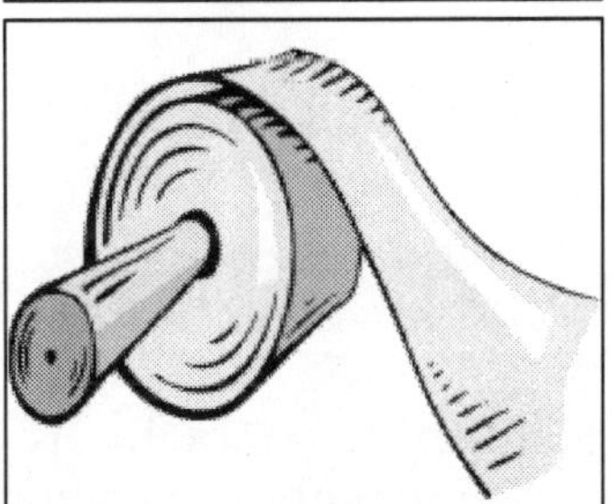

Step 02: The sheet is fed through a press which stamps out thousands of ends every minute.

Step 03: As the ends are stamped out the edges are curled at thesame time.

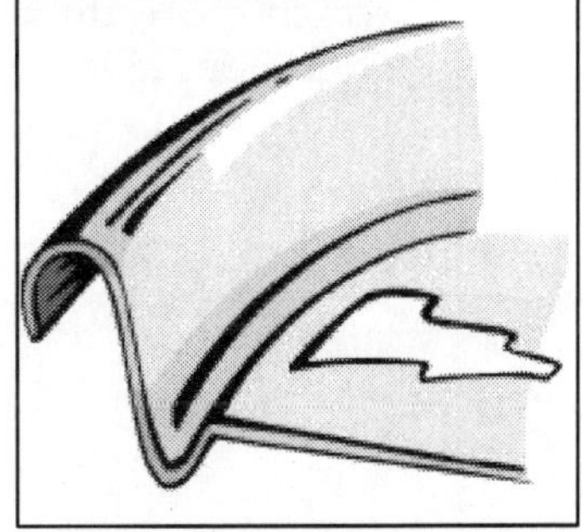

Step 04: The newly formed ends are passed through a lining machine which applies a very precise bead of compound sealant around the inside of the curl.

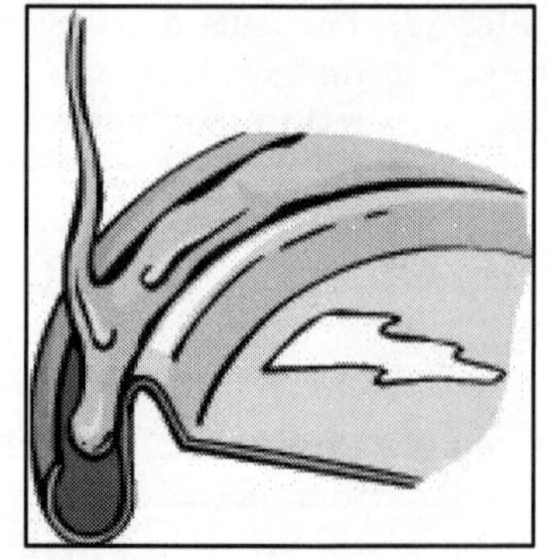

Step 05: A video inspection system checks the ends to ensure they are perfect.

Step 06: The pull tabs are made from a narrow width coil of aluminium. The strip is first pierced and cut and the tab is formed in two further stages before being joined to the can end.

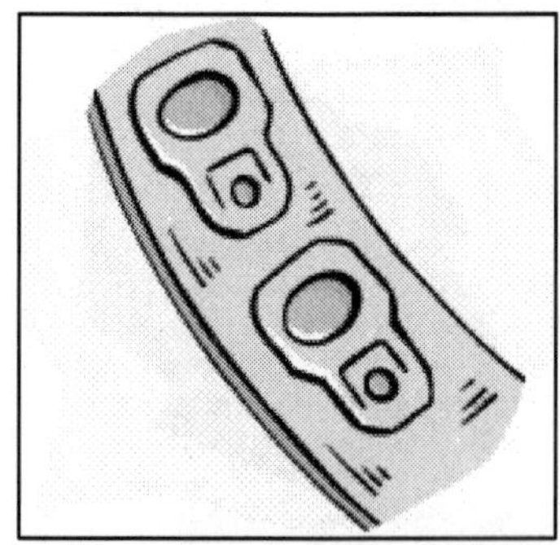

Step 07: The ends pass through a series of dies which score them and attach the tabs, which are fed in from a separate source.

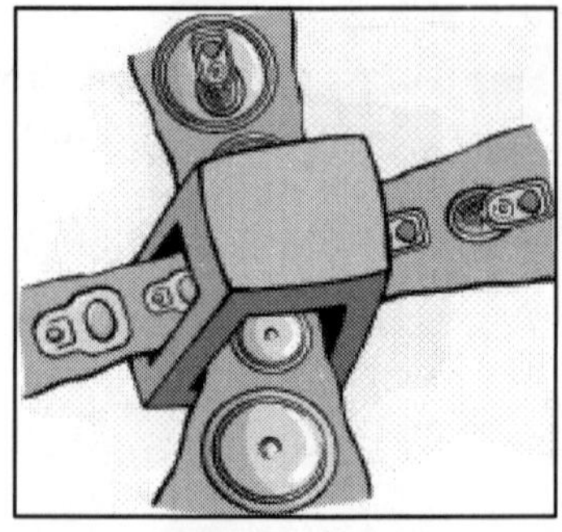

Step 08: The final product is the retained ring pull end.

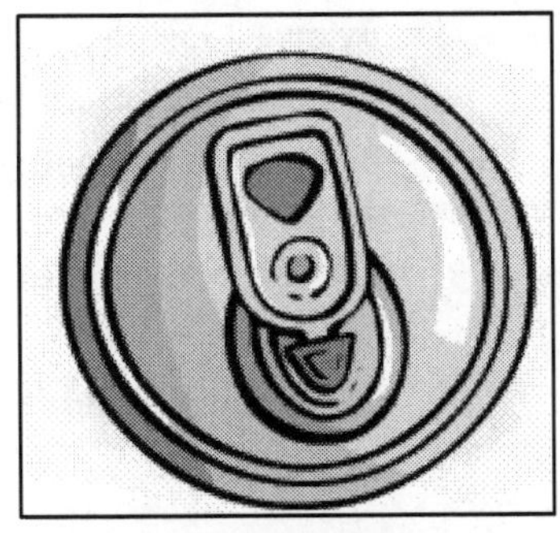

Step 09: The finished ends, ready for capping the filled cans, are packaged in paper sleeves and palletised for shipment to the can filler.

Tinplate Cans

Tinplate is the coating of TIN on to mild steel sheets. This is to protect mild steel from rusting and affecting the products packed in them.

A **tin can**, **tin** (especially in British English), **steel can**, or a **can**, is an air-tight container for the distribution or storage of goods, composed of thin metal, and requiring cutting or tearing of the metal as the means of opening. Cans hold diverse contents, but the overwhelming majority preserve food by canning. "Tin" cans are made of tinplate (tin-coated steel).

The tin can was patented in 1813 by the English inventor Peter Durand, based on experimental work by the Frenchman Nicolas Appert. He did not produce any food cans himself, but sold his patent to two other Englishmen, Bryan Donkin and John Hall, who set up a commercial canning factory, and by 1813 were producing their first canned goods for the British Army. Early cans were sealed with lead soldering, which has led to lead poisoning. Famously, in the 1845 Arctic expedition of Sir John Franklin, crew members suffered from severe lead poisoning after three years of eating canned food.

In 1901, the American Can Company was founded which, at the time, produced 90 per cent of United States tin cans.

Nomenclature of Cans

Cans can be further classified as follows:

- By shape

- By size
- By construction of can body
- By types of top components.

Classification by Shape

- Some common shapes are: round, rectangular, square, oval etc.
- Each shape requires separate production lines.
- General line cans be in different shapes.
- OTS cans are normally in rectangular shape.

Classification by Size

- Can sizes range from 50ml to 20 ltrs
- Small size cans are used in food products and in retail packs of oil, paints etc.

Food Cans

- Food cans are also called open top sanitary cans. (OTS cans).
- Used for thermally processed food packs like canned vegetables, fruits, fish, poultry, meat etc.
- These cans are distinct from general line cans.
- The quality of tin plate and the process of manufacture are rigid from raw material stage till it is used by the canner.
- OTS cans have differential coating of tin and is totally absent of defects.
- OTS cans require special understanding of the products and processes of the item being packed.
- No internal corossion due to product, atmospheric and storage conditions.
- Cans have to with stand long shelf life.
- It should be capable of with standing high speed packing and hence dimensional accuracy is critical.
- No compromise on quality.

Advantages of Tin Cans

- Stands rigours of transportation and handling
- Perfect shield against pests and rodents.
- Unlike glass can with stand pasturization and auto claving
- Ideal medium for packing processed foods
- Easy to handle and free from adulteration
- Preserves flavour and product till it reaches the consumer

- Lends itself to excellent lithography
- Easy opening and reclosing features.

Drawn and Ironed Cans

- Normally used for production of 2 piece cans.
- P The cup is first drawn and subjected to redraw till the desired diamensional

Advantages are

- High material economy
- Thin walled cans
- Saving on welding wire and lining compound
- Lower transportation cost
- Better stackability

Tinplate is essentially a blackplate coated with tin. Black plate is low carbon mild steel with a coating of oil.

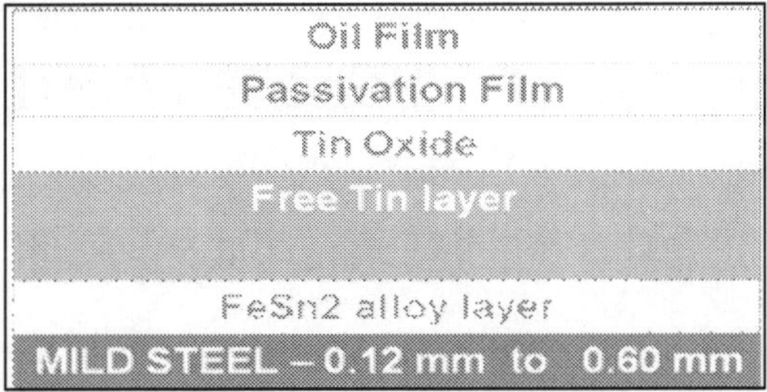

The process of making tinplate consist of the following 3 stages.

1. Hot Rolled Coil manufacture
2. Cold Rolling to produce TMBP Coils and
3. Electro – tinplating.

Hot Rolled Coil Manufacture

1. Once the steel is produced, the continuous casting process transforms molten steel into slabs in one continuous operation.
2. Continuous casting is a high energy and cost efficient process that produced a slab of excellent quality and consistency.
3. Continuous casting methods are now used invirtually all carbon steel production facilities to produce products of consistent, high quality.
4. Next, the slab goes into the Hot Strip Mill (HSM) where its thickness is reduced and it wound in a down coiler into coil form.

This forms the basic raw material of tinplate making.

Cold Rolling

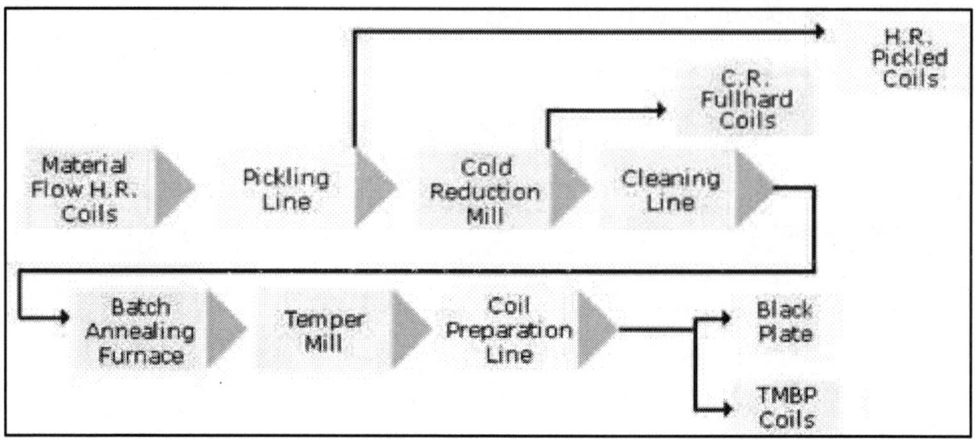

Electrolytic Tinplating Line

The unique properties of tin provide a coating on the steel that protects the contents. While providing the steel with an attractive appearance, corrosion resistance and ease in bonding wielding and paining. Tin is applied to both sides of black plate coil through an electrolytic process. The thickness of tin according to end-use applicant.

The technological advances had led to the introduction of Electrolytic Tinplate (ETP) while the spiraling and prohibitive cost of tin resulted in the manufacture of suitable substitute, namely Tin Free Steel (TFS), Chromium Coated Steel.

Flow Diagram of Electrolytic Tinplating Plants

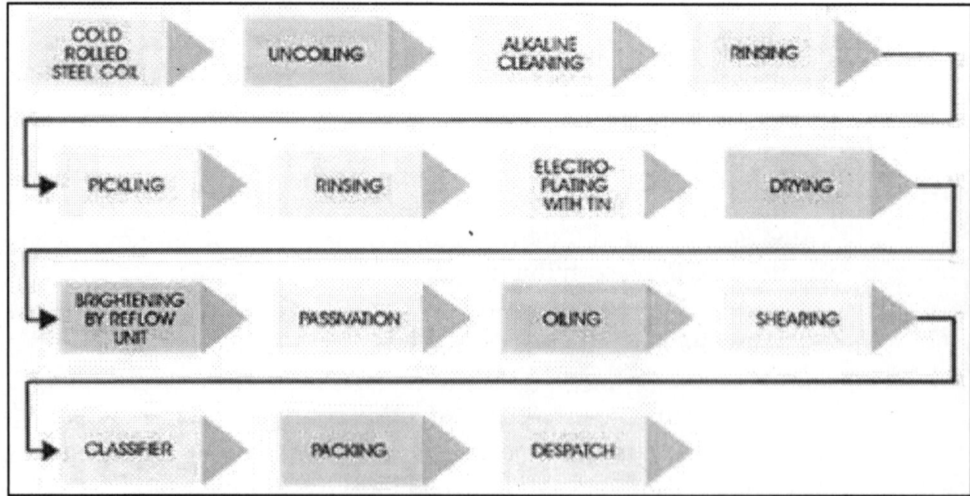

Commercial electro tinning processes are basically of two types : acid and alkaline.

Acid process requires less power hence the acid process is the most preferred method of tinplate making.

Sizing

Application of a thin coat of a synthetic resin to the sheets and curing them in an oven at elevated temperature. This imparts adhesion of further decoration to the surface of tinplate and making it suitable for fabrication under the stress of cooling.

Coating

Application of esters with pigments to sized sheets. This provides the base on background for further printing. The sheets are cured in oven after application of the coating.

Printing

Printing is done by the offset process. There can be a number of passes for printing depending on number of colours of the combined effect of various colours in the designed concerned. After each pass the sheets are cured.

Varnishing

Application of a coat of varnish and curing it in order to:

1. Impart more resistance in the decoration.
2. Impart gloss to the printed sheets.
3. Make the decoration chemically resistance.
4. Prepare it for tooling.

Lacquering

Lacquers are on the surface of sheets to :

1. Make it compatible with the product when applied on the inside surface.
2. Impart golden or similar appearance when applied on the outside.
3. Impart chemical resistance to the surface.

Manufacture in Press Shop

Slitting

Cutting of sheets into strips on Rotary slitters or scroll shears, so as to make the material suitable to be fed into presses.

Component/end manufacture on presses.

Components are made on power presses by using suitable dies within are set on them.

Ancillary Operations

These consists of either assembly of various components when required *i.e.* application of latex rubber on ends to serve as a gasket in a seam.

Manufacture of Assembly Lines

Slitting of tinplate into individual body blanks on Rotary or scroll slitters.

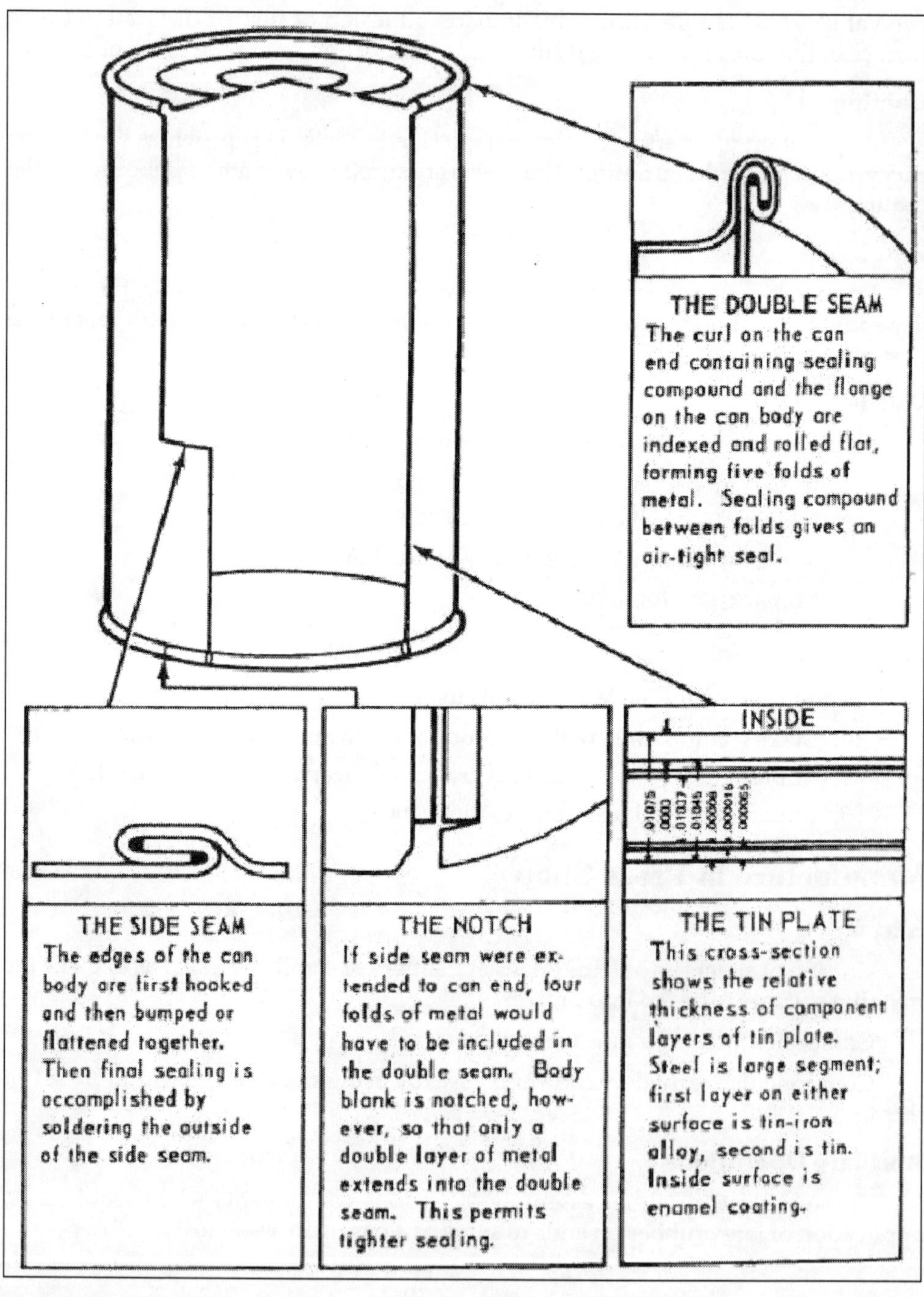

Notching

Cutting of 4 corners of body blanks to avoid excess materials at the juncture of side seam and end seam, which can lead to deformation of seam and resultant leakers.

Foldering

Folding of vertical edges of a body blank to form outer and inner hooks.

Forming

Forming the flat body blank into the corresponding shape of the can either by rolling or by forming over a mandrel.

Locking

Engaging the hooks and pressing them together to form aside seam.

Soldering/Cementing

Filling the gaps in a lock seam with molten solder or side seam cements to make the side seam leak proof.

Flanging

Flaring the open ends of the locked body so as to make body hooks which are required in formation of double seams.

End Seaming

Curling of end flanged and body flanged together and pressing them in controlled fashion and double seamers in order to produce leak proof seam.

Ancillary Operation

There are operations such as handle fitting, lug welding which are required in some cases.

Packing/Palletising

This is done manually or depending upon requirement, mode of packing being string bundles, paper bundles or palletized loads.

Galvanized and Mild Steel Drums

Galvanization is a process of coating Zinc on to the surface of Mild Steel (Iron). Generally, Galvanized and Mild Steel sheets are used to make bulk drums. Since this is not used commonly in Food industry details are beyond the scope of this syllabus.

5

Plastic

Introduction

Plastics are the most widely used material in the world now. Everywhere we look right from a watch to furniture like a chair, are now made of plastics. Plastics also have a huge performance in the packaging industry. This is because of the property if plastics to repel water and its resistance to most chemicals. Hence to understand the science in packaging, it has now become mandatory to understand the basics of plastics and their properties, and a few terms involved with the same.

What are Polymers?

Polymers are naturally occurring or synthetically produced large chain compound of repeated units called monomers. The definition of monomers is slightly recursive. Monomers are the smaller single units that make up a polymer. The number of these monomers is more than 5000 for a single polymer. They can be classified as follows:

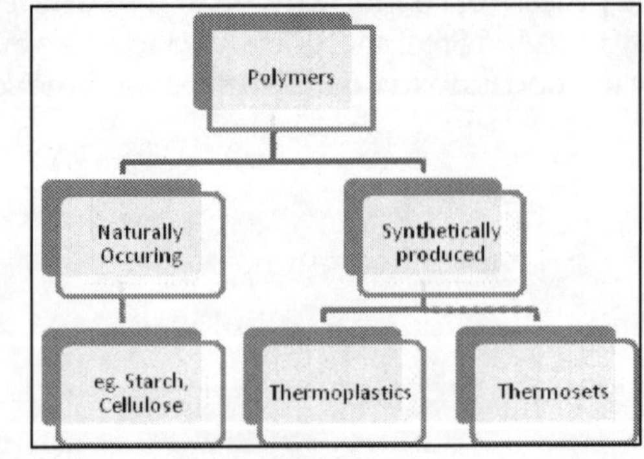

Only the synthetically produced polymers are used for packaging, especially the thermoplastics.

Polymerization

Polymerization is the reaction that which a polymer is produced. The Polymerization Reaction is of two types:

1. Addition Polymerization
2. Condensation polymerization

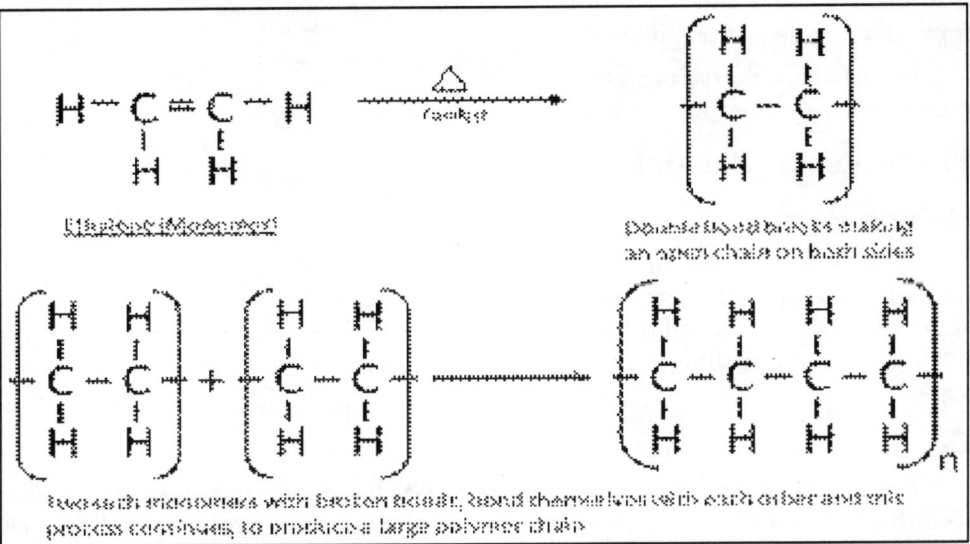

Addition Polymerization

In this process, the reacting monomers are added to each end increasing the chain to constitute the polymer. See the diagram below to understand the production of Polyethylene which is an example to addition polymerization.

Condensation Polymerization

In this process, a molecule is condensed during the reaction as a by-product or residue. See below to understand the formation of urea formaldehyde resin. In this reaction, a water molecule is thrown out to produce the Urea formaldehyde compound. The compound can again react with a Urea Molecule to increase the chain thereby forming the polymer.

Theremoplastics

Thermoplastics are polymers that are capable of remoulding. These polymers can be softend and moulded any number of times. *e.g.* Polyethylene, Polypropylene, etc. These polymers **generally** are formed by addtion polymerziation.

Thermosets

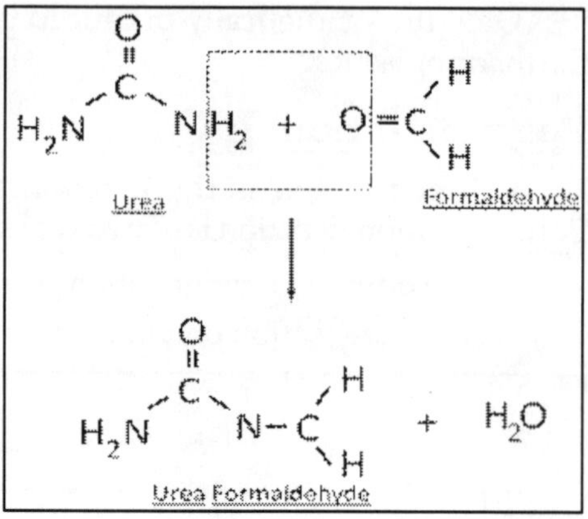

Thermosets are sythetically produced polymers that can be only moulded once. This is because they solidify on heating unlike thermoplastics which soften because of heat. Hence, if heated further more after solidification, then the Polymer disintegrates. *e.g.* Bakelite, Ureaformall-dehyde. These polymers are **generally** formed by condensation polymerization

In Packaging, the use of thermoplasics is comparitively more.

Properties of Plastics

Melting Point

Plastics do not have a melting point, they have a melting range. This is because iof their structure. Plastics do not have a specific structure. They are crystalline at some places as well as amorphous at some places. This is the reason that plastics do not have a specific melting point *e.g.:* The melting point of Polythylene is around 105–130°C.

Crystallinity

Crystallinity is the Property by virtue of which a strand of polmer can arrange itself in a crystal structure. A plastic can never be 100 per cent crystalline. The percentage of crystallinity is directly proportional to the density. If a plastic is say 40 per cent crystalline the the rest 60 per cent of it is amorphous.

Density

This is another imporant property to consider while we see the material for packaging. Density of a polymer determines its melting range, crystallinity, barrier properties and many more. The density of plastics is an important consideration in a huge number of calcutaions as any polymer always shows consistancy in its density.

Melf Flow Index (MFI)

MFI is a similar property that is to be considered whilst choosing a plastic for moulding. This is similar to the viscosity of fluids. It is defined as "The time taken by a unit quantity of polymer of flow through unit volume at its softening range". What needs to be understood is that more the MFI less viscous it will be that is more is the flow rate and vice versa.

The table below shows the raltion between the properties.

Density	*Crystallinity*	*Melt Flow Index (MFI)*
Increases	Increases	Less
Decreases	Decreases	More

Next comes the optical properties

Gloss

The amount of light that is reflected from the surface of the plastic is called Gloss. It is measured in Percentage.

Opacity

The property by virtue of which the plastic does not allow light to pass through it is called opacity. It is calso measured percentage.

Optical properties are considered during the asthetics of the package.

That completes with the basics of polemers and plastics

Plastic Processing

Introduction

Plastics are an indispensible part of our lives. We cannot directly use raw plastics and hence there are methods of processing it to suit our necessity. There are many methods and the popular methods are:

- Extrusion moulding
- Injection moulding
- Blow moulding
- Rotational moulding
- Thermoforming

1. Extrusion Moulding

This is a basic form of moulding which removes a continuous wave of molded plastic from its die. This involves a machine called Extruder. The Schematic diagram of the machine is shown below.

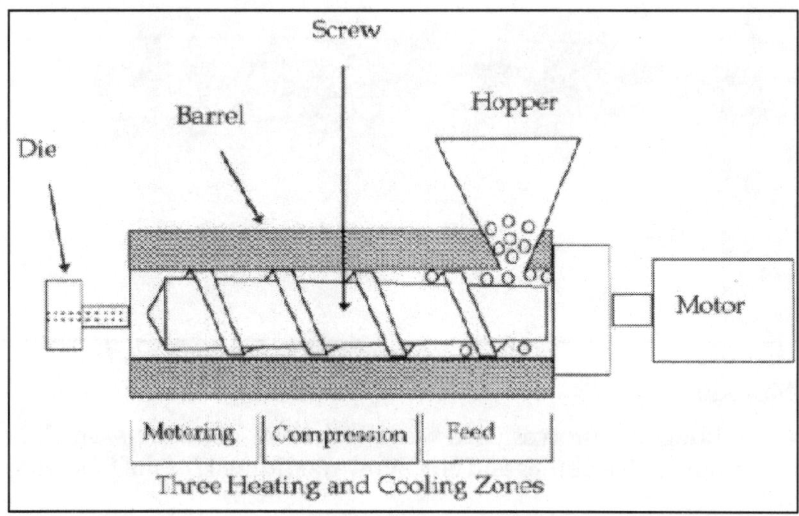

The material is fed through the hopper along with any additives necessary. The motor rotates the screw, which in turn pushes the material forward and the heating barrel in the mean time heating it. In the compression section the screw compresses the plastic and thereby mixes the plastic with its additives properly and homogenizes it. Because of the helical structure of the screw, the material is continuously pushed forward. By the time the material reaches the metering section it is a complete homogenous mixture. Then the plastic is extruded out of the die. The plastic takes the shape of the die.

2. Injection Moulding

Injection moulding is similar to the extrusion moulding, except that it is an intermittent process. See the schematic diagram below to understand the process.

The Plastic granules are fed in through the hopper. The screw, just like the extruder is continuously rotating and pushing the material forward. The barrel heats the material and homogenizes it. At the end of the machine, just before the nozzle is the injection chamber. When the enough material is accumulated in the chamber, the screw acts as a "Battering Ram" and pushes the material in a jerk out of the nozzle. The material according to the path flows into the mould and is immediately cooled, hence producing the desired object.

Injection moulding machines are of two types:

- Ram injector
- Reciprocating screw.

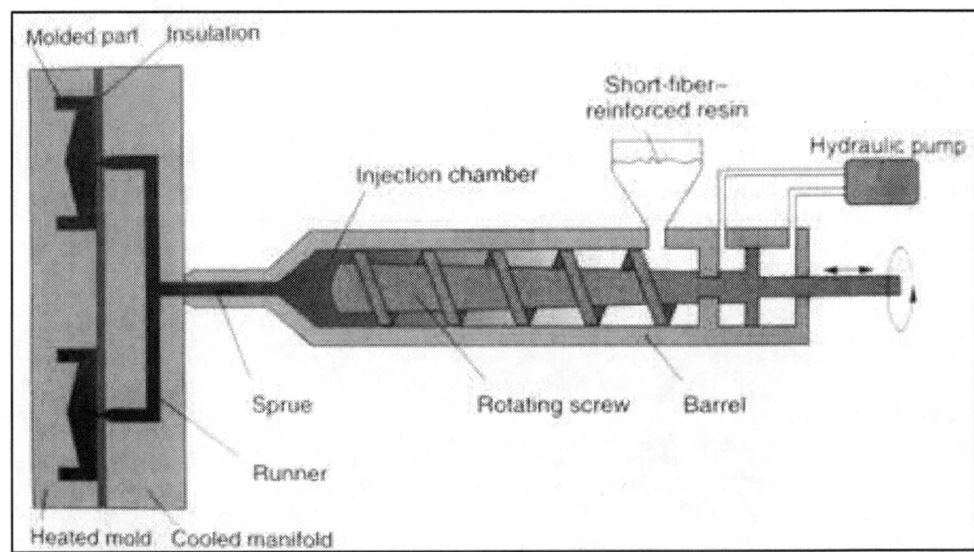

3. Blow Moulding

Blow moulding is a process used to make hollow containers, small in shape with narrow mouths, like bottles and jars. Blow moulding is of the following types:

- Injection Blow
- Extrusion Blow
- Stretch Blow

See the diagram to understand the extrusion blow moulding process.

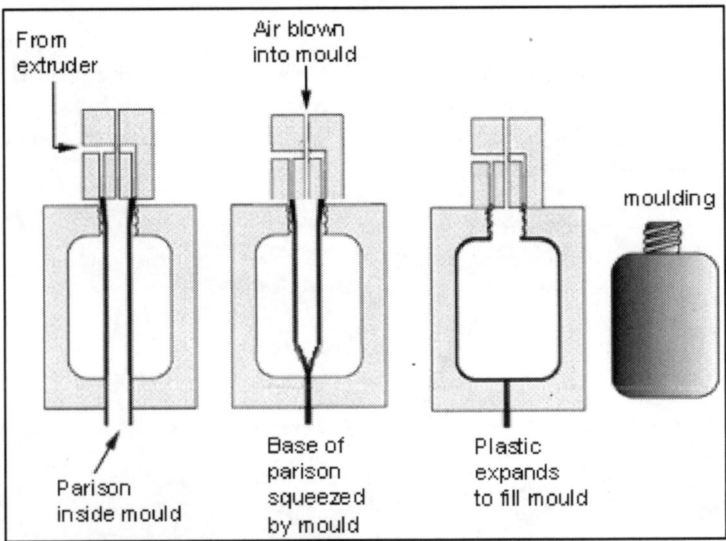

As it is seen the Parison comes out of the extruder and goes into the mould. The mould then closes squeezing the base of the parison. The air is blown in the mould. The host plastic expands and takes the shape of the mould. The mould opens and the object is removed.

The stretch blow moulding involves the use of a mandrel where the mandrel pushes down the parison stretching it to the bottom and during this process, air is blown and the parison is forced to take the shape of the container. The advantage of

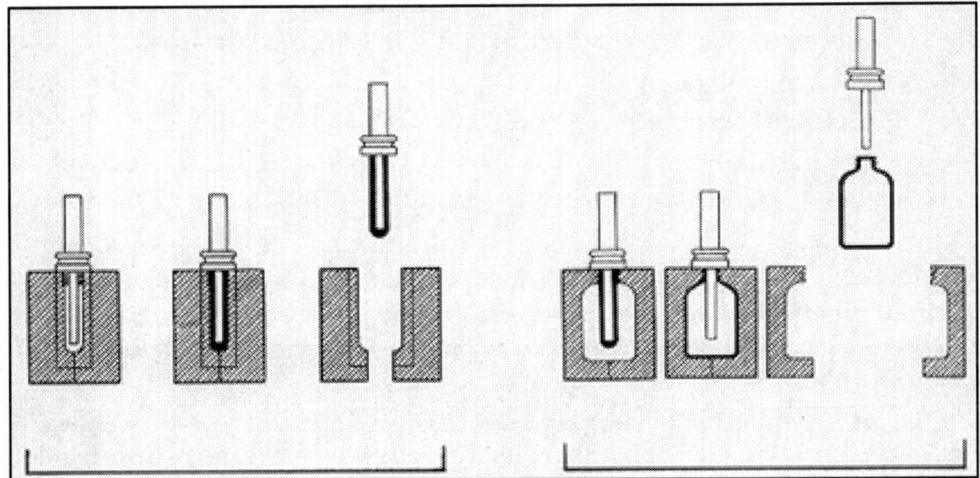

this process is that the material remains even throughout the walls of the object and the object is stress free.

4. Rotational Moulding

This process used for the manufacture of medium-large hollow objects, especially overhead tanks, used for storage of water.

Here in this process the mould is directly filled with the plastic granules and the moulding is sent into a furnace. The furnaces is closed and the mould is biaxially rotated. The plasic slowly softens, but not completely and slowly takes the shape of the mould. The mould is the slowly cooled to avoid stresses in the object. Then the object is removed and is ready for use.

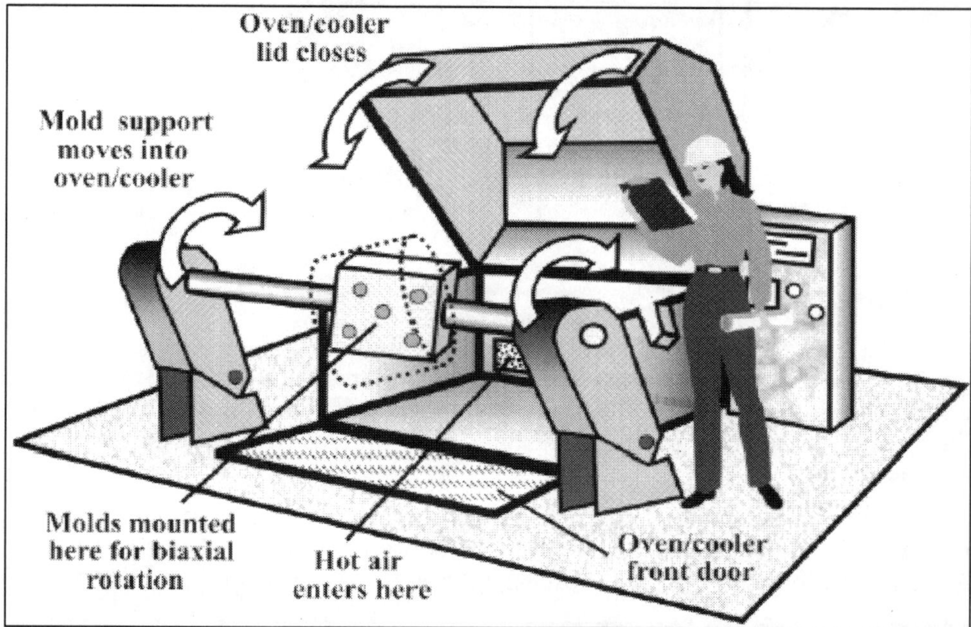

There are many types of Rotational moulding machines are few are:

- Rock and roll Machines
- Clamshell Machines
- Vertical Machines
- Shuttle Machines
- Carousel Machines

5. Thermoforming

Thermoforming is a comparatively new process used generally for the making of simple shapes.

As seen, a plastic film roll is first passed through a preheating section. Then the semi-softened plastic goes through a moulding process where there is a male mould, a female mould, and the male plunges into the female part through the softened plastic giving it the appropriate required shape. Then the edges are trimmed, making the object ready for delivery. There are 3 main methods of thermoforming. They are:

- Mechanical
- Vacuum
- Pressure thermoforming

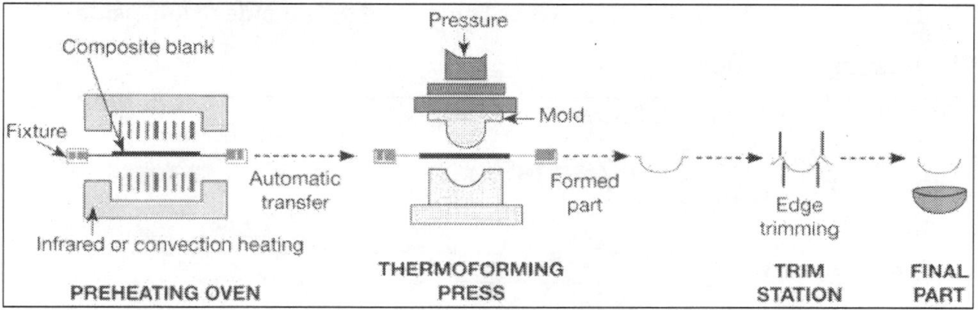

Lamintes

A product made by bonding together two or more materials. There is no perfect, universal packaging material. Laminates assemble materials with individually desirable properties to create an optimum material

Laminte Properties

Mechanical Properties

- Tensile strength
- Stiffness
- Coefficient of friction
- Use temperatures
- Elongation
- Formability

Barrier Properties

- Water vapor barrier
- Oxygen barrier
- Essential oil barrier
- Light barrier

Sealability

- Most flexible packaging is heat sealed
- Most heat seals are polyethylene based
- Other polymers used more critical applications

Aesthetic Appearance

- Clarity
- Surface gloss
- Reflective metallics

Describing Packaging Laminate

Packaging laminate plies are always listed from the outside to the inside

Plasticating Extruder

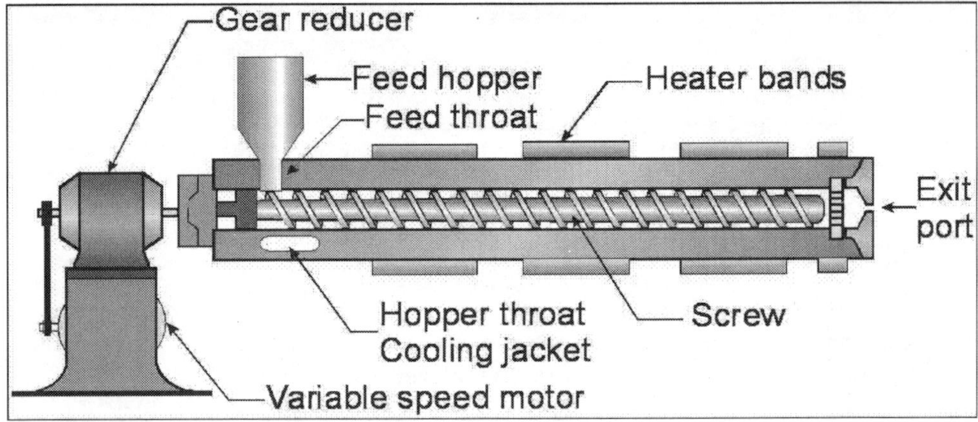

Three-Layer Coextrusion

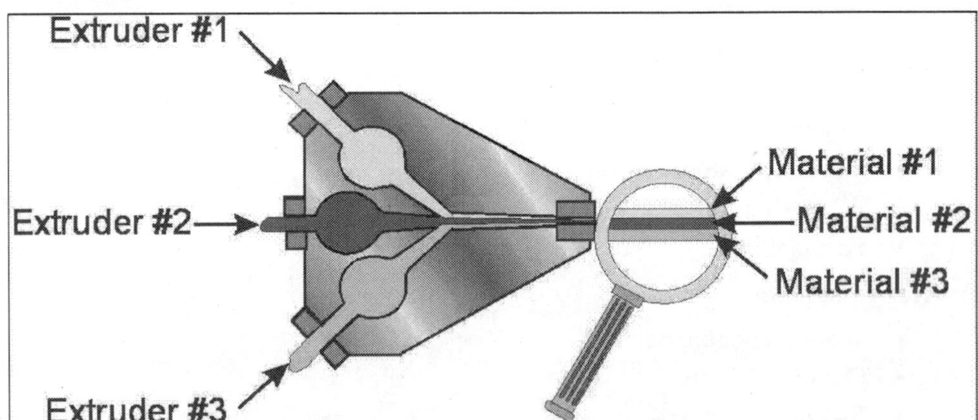

Aluminum Foil Characteristics

- Intact foil is a 100 per cent barrier to all gases
- Best deadfold properties
- Easily punctured (tamper evidence)
- Reflective of radiant heat

- Conductive (induction sealing)
- Decorative appeal: all reflective metallics are aluminum

Aluminum Foil

- Household foil is typically 17.5 mm (0.0007 inches)
- Available in gauges as low as 7 mm (0.00028 inches)
- Pin holing is present below 12 mm (0.0005 inches)
- Foil is susceptible to flex cracking
- Most foils are supported with plastic and / or paper
- Unsupported foil used for some lid-stock and tablet
- Push-through packaging

Foil Packaging Applications

Unsupported foil, non-sealable
➤ e.g. confection and cheese wraps

Unsupported foil, heat sealable
➤ e.g. lidding stock, pharmaceutical tablet backing material

Supported foil, non heat sealable
➤ e.g. decorative wraps, label stock

Supported foil, heat sealable
➤ e.g. high barrier pouches and sachets lidding stock, retort pouches

Aluminum Vacuum Metallizing

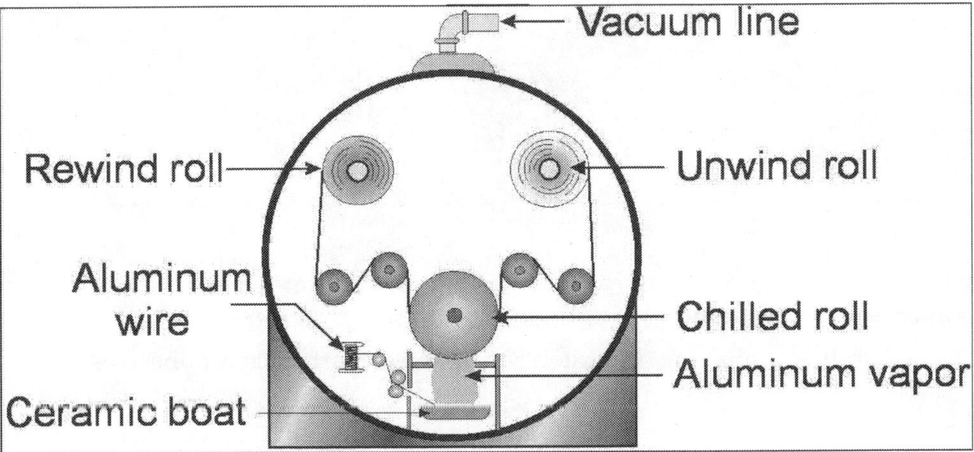

Aluminum Metallized Film

- Provides reflective metallic appearance
- Improves moisture, aroma, light and oxygen barrier
- Oxygen barrier improved: up to fifty times for OPP, up to ten times for PET
- Static dissipative applications
- OPP, PET, and PA (nylon) most common packaging films

Gravure Coating

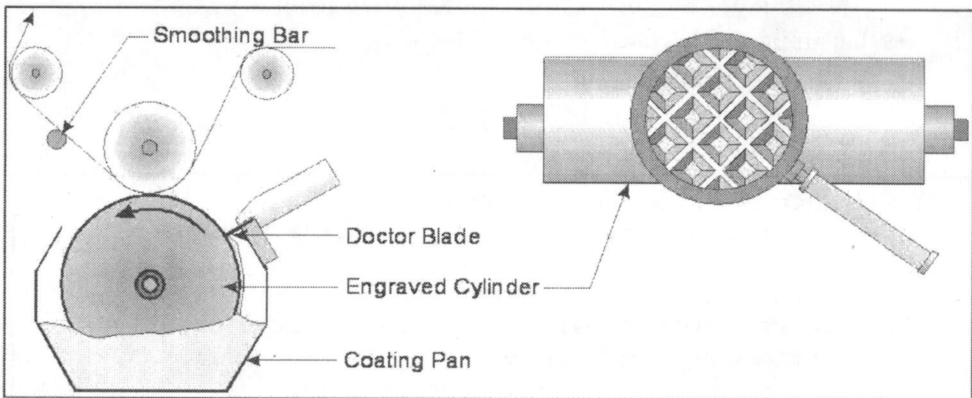

Wet Bond Laminator

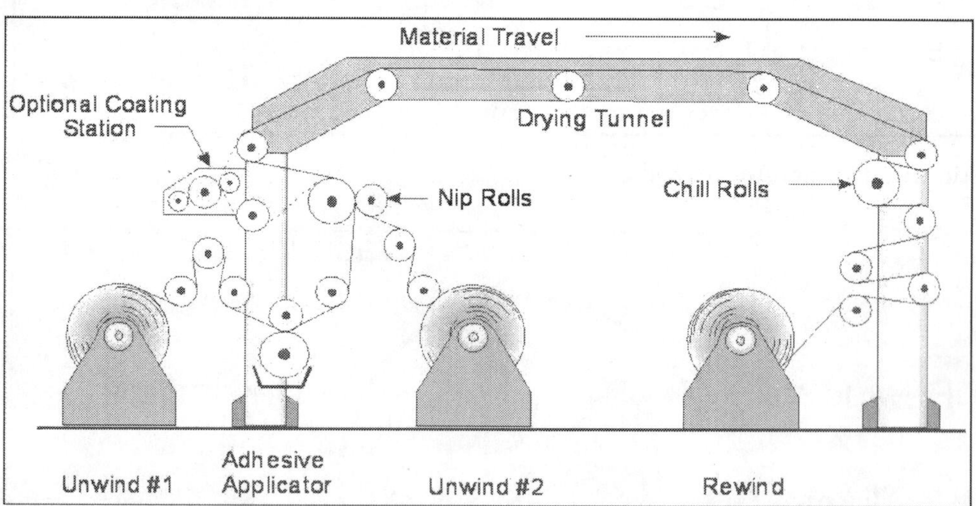

Solventless Laminating

Solventless laminating uses catalyzed or or two-part reactive adhesives

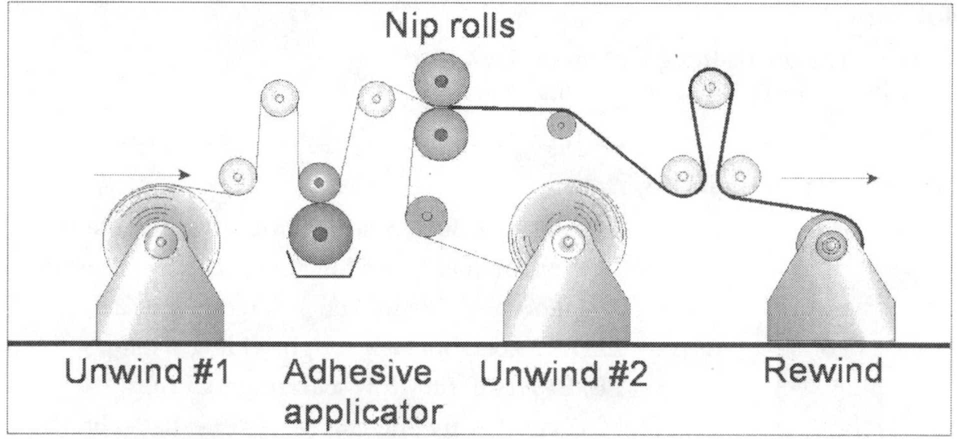

Dry Bond Laminating

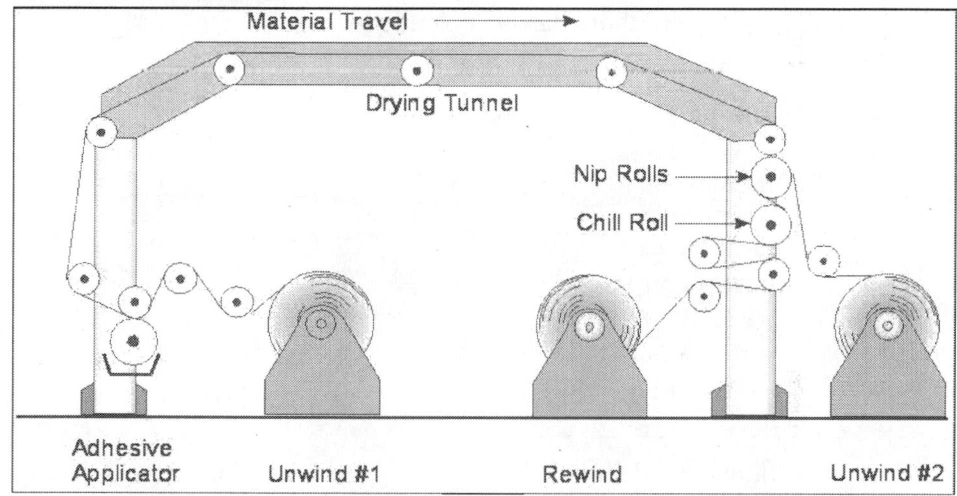

Extrusion Laminating

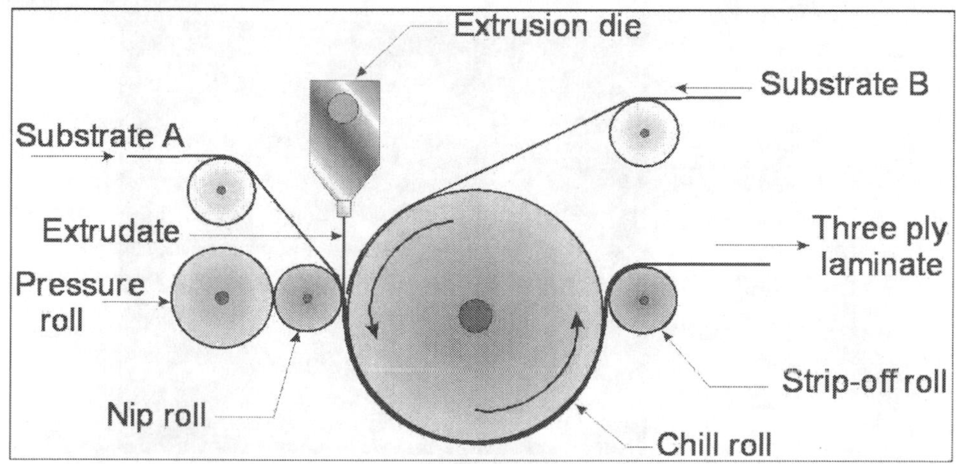

Hot Tack

- The bond strength of the seal while still hot
- Critical for most form-fill-seal machines
- Determines how quickly product can be dropped into a pouch

Common Heat- Seal Materials

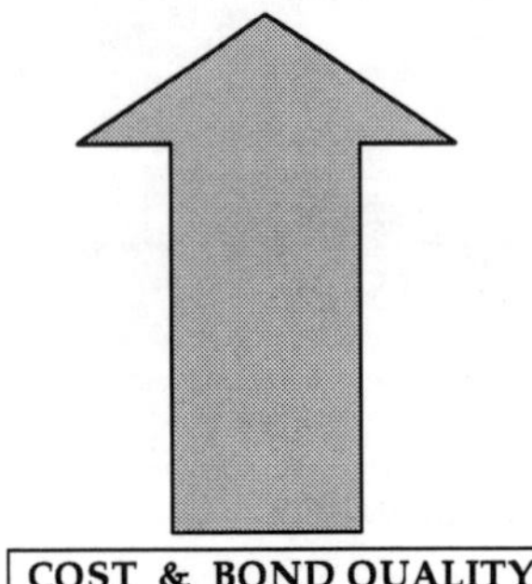

- Ionomer (*e.g.* Surlyn, seals through contaminants)
- Acid copolymer (good foil bond, chemical resistance)
- Metallocene polyethylene (low temp, fast seal)
- LLDPE (good hot tack, tough, wide seal temp.)
- PE/EVA (soft film, low seal temperature)
- Medium-density polyethylene (stiffer, better barrier)
- Cast polypropylene (stands higher temperatures)
- Low-density polyethylene

Example Laminations

REVERSE PRINTED PET
LDPE OR ADHESIVE
ALUMINIUM FOIL
LDPE OR ADHEISIVE

RETORT POUCH

REVERSE PRINTED PP
LDPE OR ADHESIVE
METALLIZED PP
LDPE SEALANT LAYER

SNACK FOOD BAG

LDPE PROTECTIVE LAYER
PRINTED PAPERBOARD
LDPE FOR BONDING
ALUMINIUM FOIL
LDPE BONDING
LDPE SEALING LAYER

ASEPTIC BOX

Laminated Collapsible Tube Construction

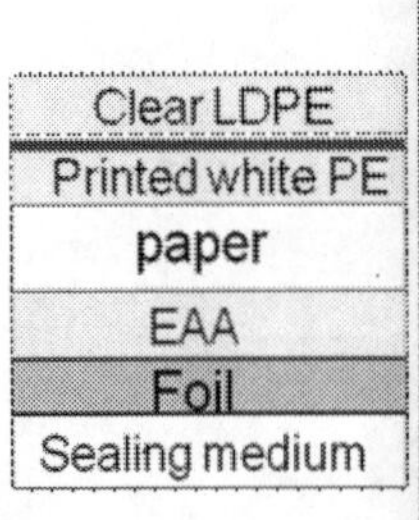

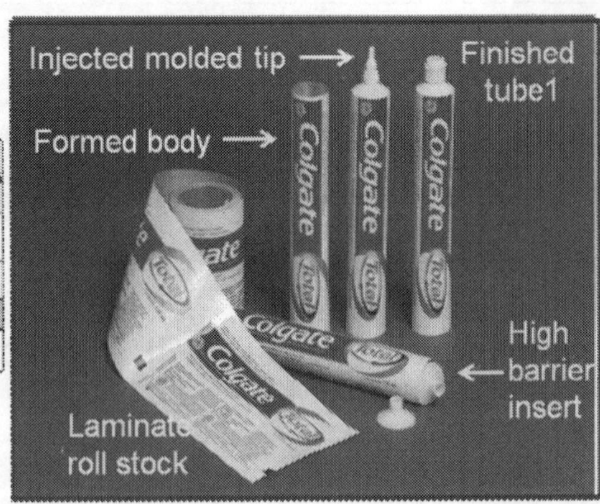

Vertical Form-Fill-Seal (VFFS)

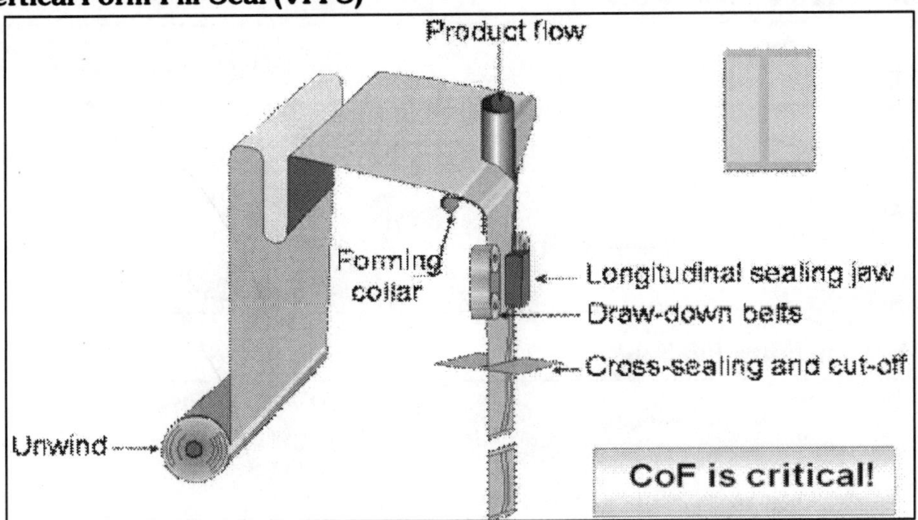

VFFS Pouch Seals Compared

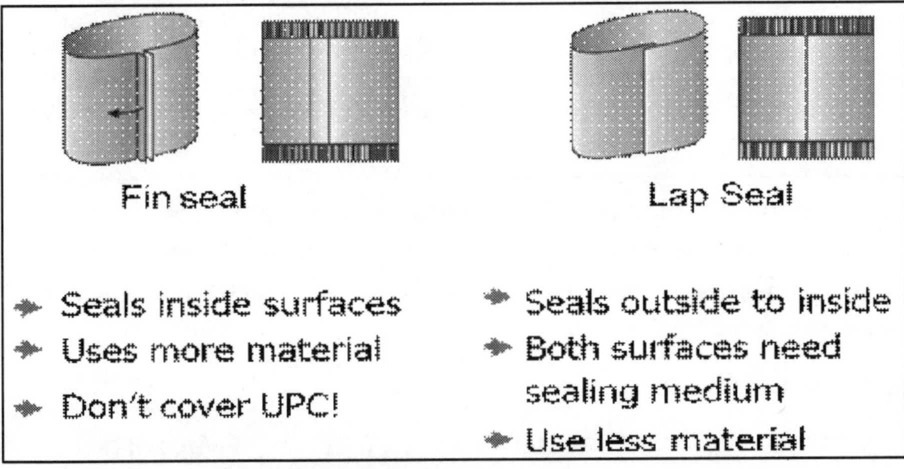

Horizontal Form-Fill-Seal (HFFS)

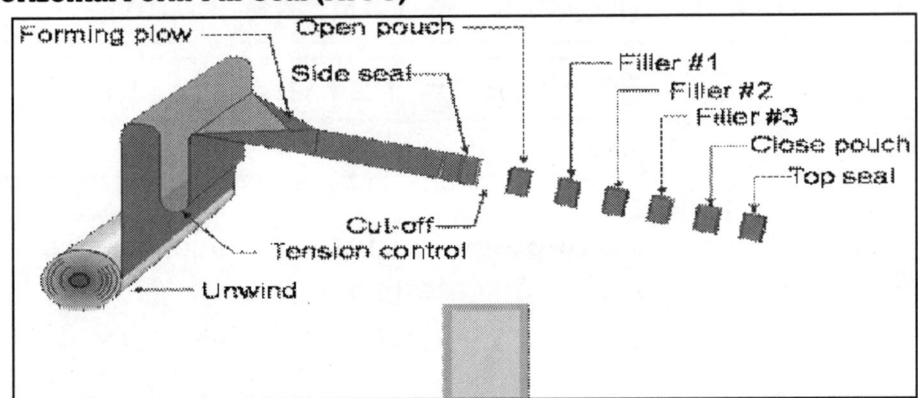

Common HFFS Pouch Styles

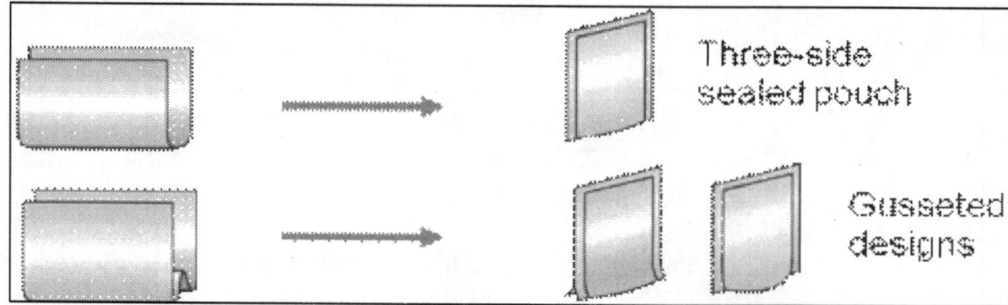

Multi-Lane Form-Fill-Seal

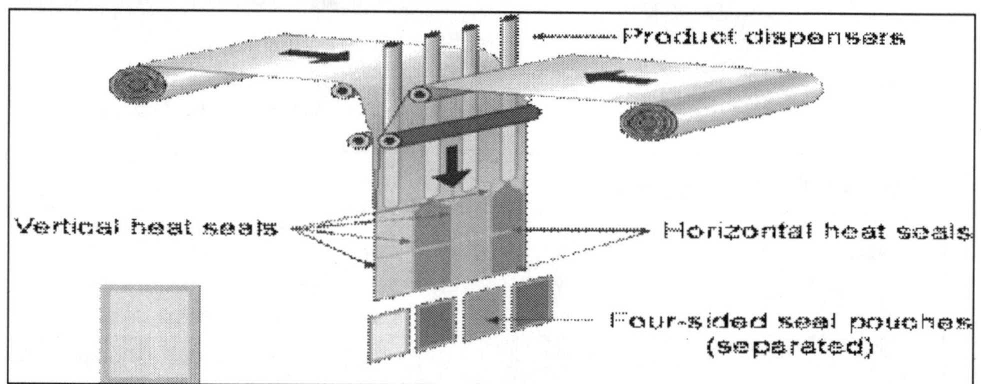

Conventional pouch has three-side seal

Can be made into various stand-up designs

Some stand-up pouches are pre-made in a separate operation

Film Thickness Measurement

Inch	Gauge	Mil	Micrometre
0.001inch =	100 gauge =	1 mil =	25 μm
0.0005 inch =	50 gauge =	1/2 mil =	13 μm

Early

ISO metric notes:

- "Micron" is a depreciated term for micrometre
- one inch = 25.4 millimetres and
- 0.001 inch = 25.4 micrometres

Universal Barriers

- PVDC is original high universal barrier polymers
- EVOH laminated between high moisture barrier
- polymers can also provide good universal barrier
- But not high enough barrier for critical applications
- Metallizing also produces a high universal barrier
- But aluminum metallized films are not transparent

Recent Advances in Achieving Barrier

- Multilayered constructions
  - PET/PA/PET
  - PET/EVOH/PET
- Interior coatings
  - SiO_x (glass) coating (Glaskin)
  - AlO_x aluminum oxide coating
  - Carbon coating (DLC and Actis)
- Exterior coatings and interior laminates
  - Epoxy-amine coating (Sealica and Bairocade)
  - Copolyester (Amosorb: interior layer)
- Nanocomposites
  - Extremely fine mineral additions

6

Wood, Jute and Cotton

Wood and Plywood in Packaging

A wooden box is a container made of wood for storage or shipping.

Wood is basically a strong and good raw material for packaging, provided its strength and economy is utilized correctly

Wooden boxes are often used for packing

- When high strength is needed
- For heavy and difficult loads
- Long term warehousing needed.
- Large size is required
- Rigidity and stacking strength is critical

Wooden boxes and crates are not the same. If the sheathing of the container (plywood, lumber etc) can be removed, and a framed structure will remain standing, the container would likely be termed a crate. If the sheathing is in place, the container would likely be termed a wooden box

The strength of a wooden box is rated based on the weight it can carry.

"Skids" or thick bottom runners, are sometimes specified to allow forklift trucks access for lifting

Performance is strongly influenced by the specific design, type of wood, type of fasteners(nails, straps) workmanship, etc.

This calls for a certain basic knowledge about wooden structures, assembling methods, etc., which unfortunately, is non-existent in most cases.

General Issues

- Moisture Content of wood are seldom proper
- Quality of wood used has defects like knots
- Improper treatment
- Hard wood is often used,
- Nails are not of the correct sizes, etc.

All this, unfortunately, has a disastrous effect on strength and economy of wooden packages as they are used today for transport packaging purposes.

Types of Wood Used for Packaging

In principle, there are no particular specifications for what kind of wood should be used for a specific type of package. The choice of the species to be used will be determined, above all, by the quantity available and its price. The actual strength characteristics of the package, however, are greatly related to the type of wood, its quality, its thickness, and the workmanship in the construction and assembly of the package.

In this context, different types of woods have often vastly different properties, such as:

- Density or unit weight (expressed in lbs. / cub.ft; kg / m^3)
- Bending strength (lbs. / sq.in.; kg / cm^2);
- Compressive strength (lbs. / sq.in.; kg / cm^2);
- Nail holding power
- Resistance to splitting
- Ease of working;
- Resistance to decay, etc

All wood falls into two general categories:

Softwood: which comes from coniferous or needle-bearing trees

Hardwood: which comes from deciduous, broad-leafed trees.

A number of further classifications of species have been made for packaging purposes, largely on the basis of density (400 to 750 kg / m^3) and nail holding power.

Group I

- Varieties of both softwoods and hardwoods;
- Do not split easily when nailed,
- Moderate nail-holding capacity,
- Moderate strength as a beam,

- Moderate shock resisting capacity.
- Soft, light in weight, easy to work, hold their shape, and are usually easy to dry.

Group II

- Heavier coniferous species
- Greater nail-holding capacity than Group I wood
- But are also more inclined to split.
- Hard bands of summer wood tend to deflect nails and cause them to run out at the sides of the cleats

Group III

- Hardwoods of medium density.
- Have same nail-holding capacity and strength as Group II
- But are less inclined tosplit or shatter at impacts.
- They are the most suitable woods for box ends and cleats and they are also widely used for in wirebound boxes

Group IV

- Heavy hardwood species with highest densities.
- Have greatest capacity both to resist shocks
- And to hold nails
- But because of their hardness they are difficult to nail and also have the greatest tendency to split at the nails.
- Particularly suited for load-bearing members, skids.

Density of Wood

The density of the wood is an important characteristic because it gives a good indication of the strength of the wood and its resistance to the extraction of nails. Density also indicates how much shrinkage - and hence, distortion - is likely to take place during drying. Wood with a density of over 750 kg/m3 should not be used for packaging and, on the other hand, it is not advisable to use densities of less than 400 kg/m3, since these woods will not have sufficient mechanical strength. Although very tough and resistant, wood with a density of over 750 kg/m3 has a marked tendency for distortion. It holds nails well, but they are very difficult to drive in properly as the wood splits or the nails bend. This type is also unnecessarily heavy for packaging applications and should, therefore, be avoided for use as a packaging material.

High-density woods (600 - 750 kg/m^3) are used for e.g.: edge planks and spacers of pallets; load-bearing members of crates; skids, upright and outer lengthwise members of crates; cleats and battens of nailed wooden, plywood and wirebound boxes, etc. Woods of longer density (400 - 600 kg/m3) should be used for package components that are not so susceptible to stresses, such as: Intermediate upright and lengthwise members of crates cleating material; panels of box ends and sides; wirebound and light wooden packages, etc.

The wood of living trees contains a large amount of water - green or newly sawn wood can have a moisture content of over 200 per cent. For most uses of wood as a packaging material, most of this moisture has to be removed, and the wood seasoned by air or kiln drying. The moisture content of wood is one of the principal factors affecting its strength.

Generally, as wood dries, most of its important strength characteristics increase. This increase in strength, however, does not occur until the drying has reached the fibre-saturation point, which is the condition in which the water has evaporated from the cell cavities but the cell walls are still fully saturated with water.

For practical purposes, the fibre-saturation point is considered to be approximately 30 per cent moisture for most species. Material dried to 12 per cent moisture content may be twice as strong in bending compared to green material, and if the lumber is kiln dried to 5 per cent, its bending strength may be tripled. The resistance to extraction of nails may be as much as 30 per cent greater for dry wood than for green wood. However, as wood dries its toughness and shock resistance might decrease. This is because dried wood will not bend as far as green wood before failure, although it will sustain a greater load. When the moisture of wood falls below its saturation point, the wood begins to shrink. The wood stops drying when it has reached an equilibrium with the temperature and the relative humidity conditions of the surrounding air. This equilibrium point varies from 10 - 25 per cent, depending on climatic conditions. Shrinking of a wooden package, made out of green wood, considerably weakens the strength of its construction by distortion, checking, splitting, cupping of the timber used, and reduces the holding power of nails, etc. It is therefore important to use only seasoned materials, with a moisture content never exceeding 20 per cent (12 – 18 per cent is ideal), for wooden packages. This should be well below the fibre-saturation point and close to the equilibrium point, also taking into particular consideration the climatic conditions (temperature and relative humidity) in the target markets for the export packages.

Another important point in this context, is the possible savings in freight through a reduction of shipping weights. A wooden package might have a tare weight of 20 lbs. (9 kg) when made of wood having a moisture content of 80 per cent, but if made of well-seasoned wood with *e.g.* 15 per cent moisture, the box itself would weigh approximately 13 lbs. (6 kg). The saving of 7 lbs. (3 kg) in tare weight will result in a direct saving in freight charges. A similar effect can be achieved through reducing lumber dimensions (less tare weights) for seasoned wood with strength values equal to those of previously used green wood

There are two additional disadvantages of using wood which is too moist in packaging.

- Firstly, it substantially increases risk for corrosion or moulding of contents of the package.

- Secondly, the wet wood itself is more likely to be attacked by wood destroying fungi, leading to decay, loss of strength, and possible negative side effects for the packed product inside.

Defects in Lumber

For economical reasons, it will obviously not be possible to use first-class lumber as a raw material for wooden packaging. Certain defects might therefore be allowed, but should not materially reduce the structural strength nor interfere with the most effective nailing methods or patterns. There are two types of defects which have a major impact on package strength:

(i) Cross Grain/Slat Grain

(ii) Knots

(i) Cross or Slanting Grain

The slope of the grain is the direction of the wood fibres with respect to the longitudinal axis of a piece of wood. When the fibres are not parallel to this axis, the wood is said to have a slanting grain (cross grain). The slope represented by the angle between the direction of the grain and the longitudinal axis of the piece of wood is expressed as a percentage. Slight local deviations in the grain are usually ignored. However, a grain slanting more substantially causes a considerable reduction in strength. Thus, for instance, a slope of 10 per cent causes a reduction of 40 per cent in the bending strength, and a slope of 15 per cent, a reduction of 60 per cent. Cross grain with a slope of more than 5 per cent should, therefore, be avoided. In addition to the loss of strength, cross grain wood has a marked tendency to warp when subjected to variations in humidity, and the risk for the wood splitting by nailing is increased.

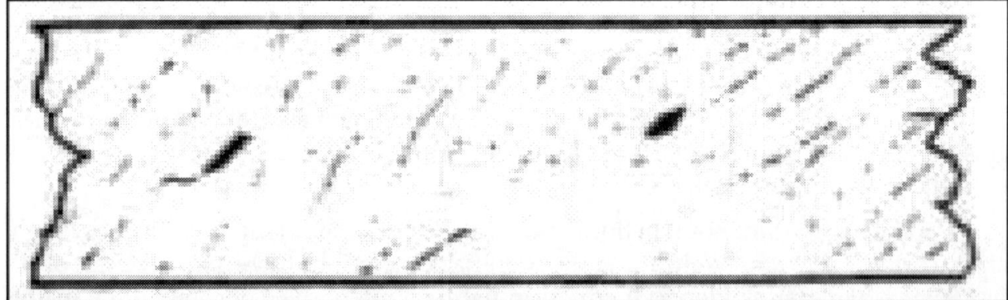

(ii) Knots

The weakening of the bending strength is nearly proportional to the diameter of the knot measured across the width of the board. This weakening effect is caused mainly by the cross grain wood around the knot. A distinction should, however, be made between healthy knots, strongly attached to the surrounding wood, and black, rotten, loose knots. Knots weaken the board most if they are in the middle third of the length of the board. At no time should the diameter of the knot exceed one third of the width of the board. Large size knots should be eliminated

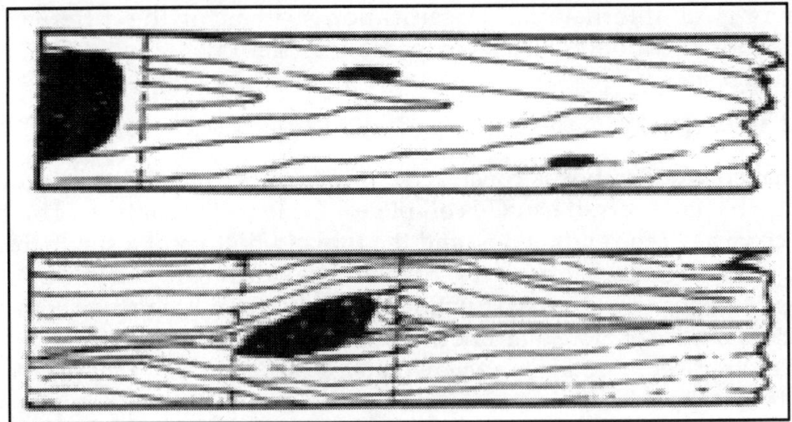

(iii) Checking

Caused by stresses from non-uniform shrinkage. End checking is caused by the wood drying more rapidly at the ends than away from the ends. This condition can often be avoided by painting the ends to reduce their drying, or by reducing the air circulation around the ends. Checking reduces the holding power of nails and sometimes results in splits, running the full length of the wooden piece.

(iv) Cupping

is an excessive curvature of lumber across the grain and might happen when one side of a board dries more rapidly than the other. This condition isusually temporary, but can be permanent if sawed timber is dried with insufficient weight on it.Improper drying techniques are also responsible for the board twisting, bowing or warping. Case or surface hardening is a condition caused by too rapid surface drying and might cause warping.

(v) Collapse

It is an abnormal type of shrinkage in certain types of lumber and makes the surface of the boards look caved-in or corrugated when dried. Most of these defects do not directly affect the strength of the package but they make fabrication more difficult.

Types of Wooden Packages

The types of packages made out of wooden materials can be classified into the following categories:

- Nailed wooden boxes
- Cleated wooden boxes
- Wirebound boxes
- Skid boxes.

There is no international classification system for the different types of constructions of wooden boxes comparable to the International Case Code for corrugated boxes.

Nailed Wood Box

A nailed wooden box is constructed of pieces of lumber attached by nails or other suitable fasteners. It usually completely encloses the contents. This basic box design consists of the ends, sides, and the top and bottom of a single thickness of lumber, made of one or more pieces of wood. The sides are nailed into the end grain of the end, which makes the construction relatively weak. When the depth of the box necessitates the use of more than one piece of wood in the sides or ends, it is desirable to have them meet in a staggered way (at least 1½ inch; 40 mm) in the corners.It is desirable that the sides and ends should each be in one piece, or joined together by "tongue and groove" method. This style finds applications as a returnable crate for beverages or as a field box for fruit and vegetables.

Cleated Box

A cleated box has five or six panel faces with wood strips attached to them. The panels can be made of plywood, solid or corrugated fibreboard etc. Wooden cleats reinforce the panels.

Wirebound Box

Very thin lumber is used for a wire bound box. Wires are stapled or stitched to the girth and to wood cleats. These are sometimes used for produce and for heavy

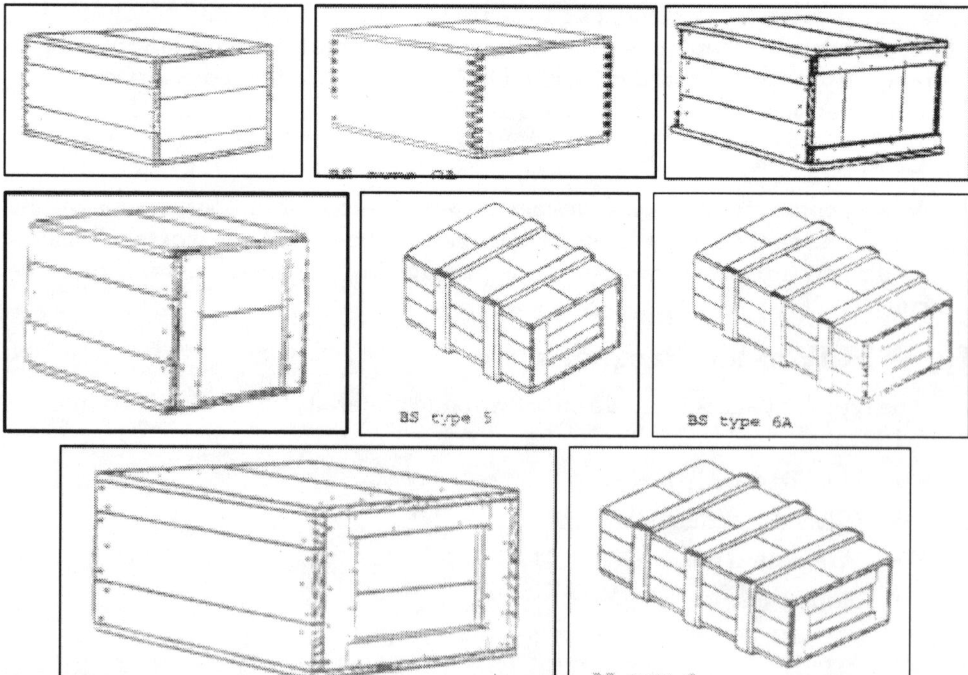

loose items for military or export use. These are lighter than wood boxes or crates. They have excellent tensile strength to contain items but not much stacking strength

Skid Box

A skid box is a wood, corrugated board or metal box attached to a heavy duty pallet or platform on a skid (parallel wood runners)

Crates

A crate is a large container, often made of wood, often used to transport large, heavy or awkward items. A crate has a self-supporting structure, with or without sheathing. For a container to be a crate, all six of its sides must be put in place to result in the rated strength of the container.

Boxes and crates are often confused with one another; mostly when they are made of wood. Contrary to a crate, the strength of a wooden box is rated based on the weight it can carry before the cap (top, ends, and sides) is installed.

Crate History

The first documented writings in the US about shipping crates is in a 1930 handbook (Technical Bulletin No. 171) written by C. A. Plaskett for the US Department of Agriculture although his writing imply that crates were defined before that time. C. A. Plaskett was known for his for his extensive testing and defining of various components of transport packaging.

Crate Construction

Although the definition of a wooden crate as compared to a wooden box is clear, construction of the two often result in a container that is not clearly a crate or a box. Both wooden crates and wooden boxes are constructed to contain unique items, the design of either a crate or box may result in the use of principles from both. In this case, the container typically will be defined by how the edges and corners of the container is constructed. If the sheathing (either plywood or lumber) can be removed, and a framed structure will remain standing, the container would likely be termed a crate.

Crate Design

There are many variations of wooden crate designs. By far the most common are 'closed', 'open' and 'framed'. A Closed Crate is one that is completely or nearly completely enclosed with material such as plywood or lumber boards. When lumber is used, gaps are often left between the boards to allow for expansion. An Open Crate is one that (typically) uses lumber for sheathing. The sheathing is typically gapped by at various distances.

There is no strict definition of an open crate as compared to a closed crate. Typically when the gap between boards is greater than the distance required for expansion, the crate would be considered an open crate. The gap between boards would typically not be greater than the width of the sheathing boards. When the gap is larger, the boards are often considered 'cleats' rather than sheathing thus rendering the crate unsheathed. An unsheathed crate is a frame crate.

A Frame Crate is one that only contains a skeletal structure and no material is added for surface or pilferage protection. Typically an open crate will be constructed of 12 pieces of lumber, each along an outer edge of the content and more lumber placed diagonally to avoid distortion from torque. When any type crate reaches a certain size, more boards may be added. These boards are often called **Cleats**. A cleat is used to provide support to a panel when that panel has reached a size that is may require added support based on the method of transportation. Cleats may be placed anywhere between the edges of a given panel.

"Skids" or thick bottom runners, are sometimes specified to allow forklift trucks access for lifting.

Transportation methods and storage conditions must always be considered when designing a crate. Every step of the transportation chain will result in different stresses from shock and vibration. Differences in pressure, temperature and humidity may not only adversely affect the content of the crate, but also will have an effect on the holding strength of the fasteners (mostly the nails and staples) in the crate.

IATA, the International Air Transport Association, for example, doesn't allow crates on airplanes because it defines a crate as an open transport container. Although a crate can be of the Open or Framed variety, having no sheathing, a Closed crate is not open and is equally as safe to ship in as a wooden box, which is allowed by IATA.

ASTM Standards

- D6039 Std Specn for Crates, Wood, Open and Covered
- D6199 Quality of Wood Members of Containers and Pallet
- D6253 Treatment and / or Marking of Wood Packaging materials
- D6251 Std Specn for Wood-Cleated Panel board Shipping Boxes
- D6254 Std Specn for Wirebound Pallet-Type Wood Boxes
- D6256 Std Specn for Wood-Cleated Shipping Boxes and Skidded, Load-Bearing Bases
- D6573 Standard specfication for General Purpose Wirebound Shipping Boxes
- D6880-05 Std Specification for wooden boxes

Construction and Nailing of Wooden Boxes/Crates

The nailing technique used in assembly and closure is one of the most important factors influencing economy and strength in the design of wooden packages. The types of nails, their sizes, the spacing and location of the nails in relation to *e.g.* the grain of the wood, greatly affect the durability of the package. It is wasteful to construct a wooden box or crate with wood having good strength characteristics and then fail to nail the parts together in an effective way. Too few nails, or nails that are too small, do not provide enough strength; likewise, nails that are too large may split the wood and weaken the construction.

Types of Nails

Dozens of different types of nails are available in market and variations and properties differ substantially from one supplier to another. Nails are classified by the primary function, special shapes, coatings, gauges, sizes and types of heads and points. Some of the more frequent designations for the nails used in packaging are:

1. Common or Bright nails
2. Bright box or Standard box nails
3. Coolers
4. Sinkers
5. Clout nails (for plywood constructions)
6. Spirally grooved nails
7. Annular grooved nails
8. Barbed nails

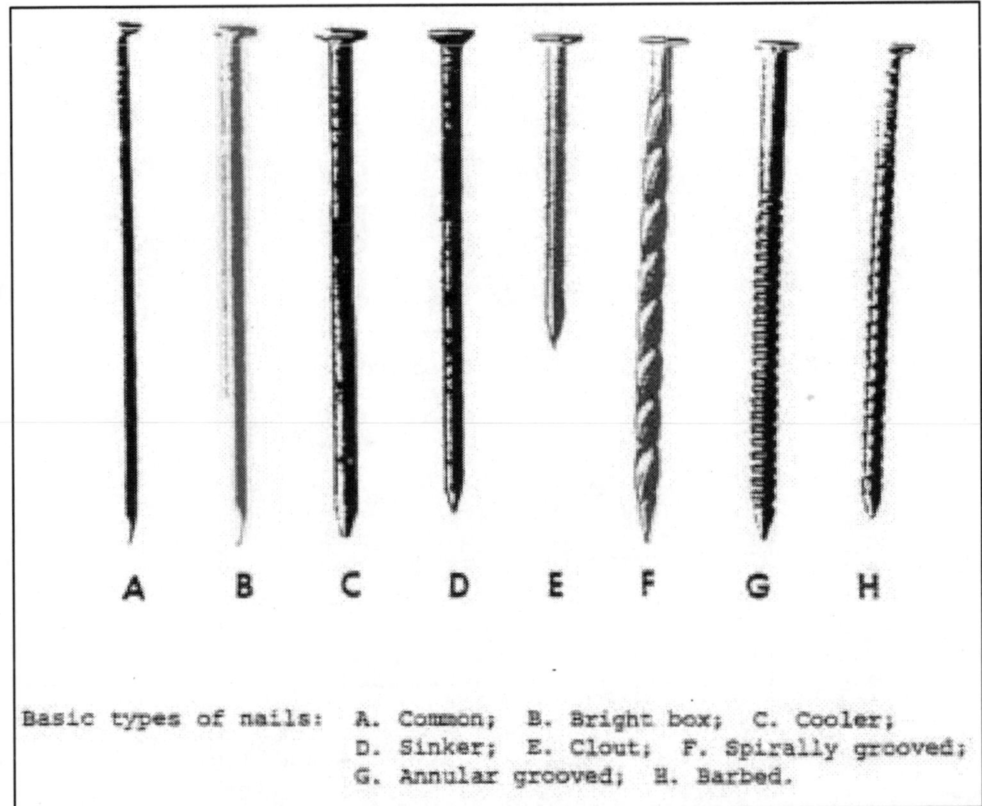

Basic types of nails: A. Common; B. Bright box; C. Cooler; D. Sinker; E. Clout; F. Spirally grooved; G. Annular grooved; H. Barbed.

If nails with a sharp point tend to split the wood, the points should be slightly blunted with the hammer before nailing. With easy-splitting hardwoods it might be necessary in some cases to drill lead holes before nailing, particularly when large size nails have to be used.

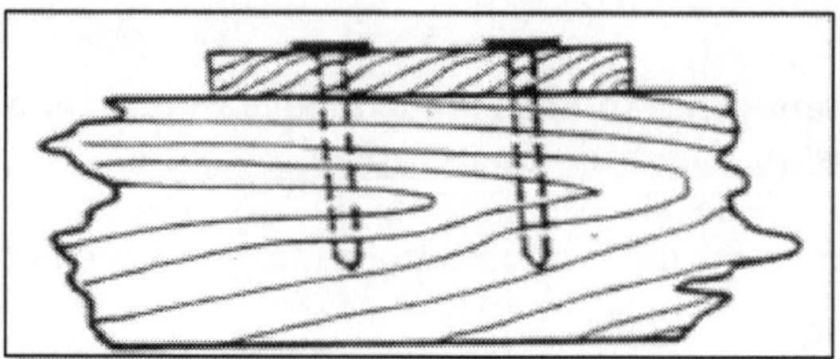

Correctly Spaced Nail

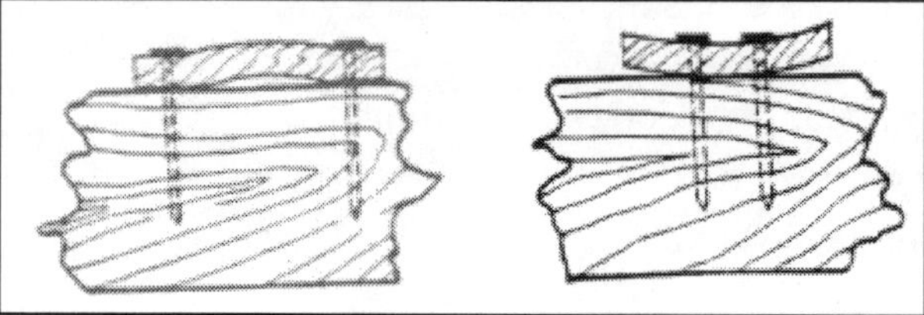

Incorrectly Spaced Nail

Splits Caused by Incorrect Nail Size

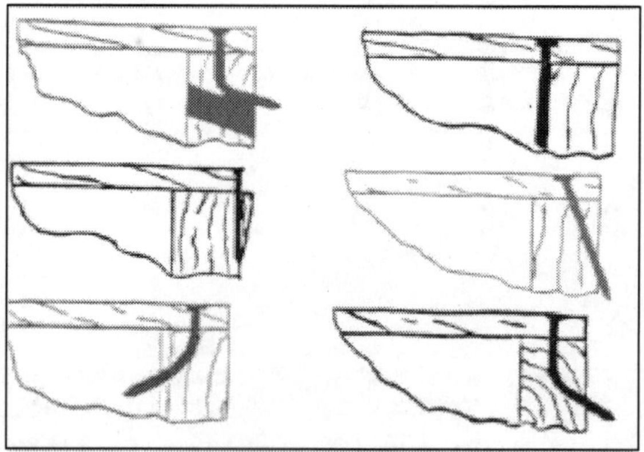

Typical Nailing Failures

Selection of Correct Timber Sizes

In this context it is impossible to give precise recommendations for the construction and selection of timber sizes for all types of nailed wooden packages.

Some of the variable factors influencing this selection are:

1. Export or domestic shipment
2. Mode of transport
3. Disposable or returnable package
4. Type of product packed: Easy load, Average load, Difficult load
5. Weight of contents
6. Type of available wood (Group I-IV)
7. Type of box and crate construction
8. Types and sizes of available nails.

7

Glass

What is Glass?

It is a *Supercooled Liquid.* It is a liquid which is cooled to a stage where its viscosity is so great that the molecules do not move freely enough to form crystals.

The first man-molded glass may have appeared around 7,000 B.C. in the form of beaded jewelry made from natural glass such as obsidian, rock crystal, agate, or onyx. It wasn't until 1,500 B.C. that the first glass containers were produced. Thousands of years later, in 1608, America's glass container industry was born when English settlers in Jamestown, Virginia, built a glass-melting furnace. By 1880, more than 25 percent of the glass made in America was used for common bottles. Glass containers remained something of a luxury, with the industry dependent on the skill and availability of its glassblowers. In 1903, all that changed with Michael J. Owens' invention of the first completely automatic glass bottle-blowing machine. This machine made it possible to mass-produce bottles and jars of uniform height, weight, and capacity. High-speed filling and packing lines soon followed. Glass containers entered the modern age, with today's machines capable of producing over 1,000,000 bottles a day!

What is Glass made of?

Sand – 70 per cent

Soda Ash – 15 per cent

Limestone – 10 per cent

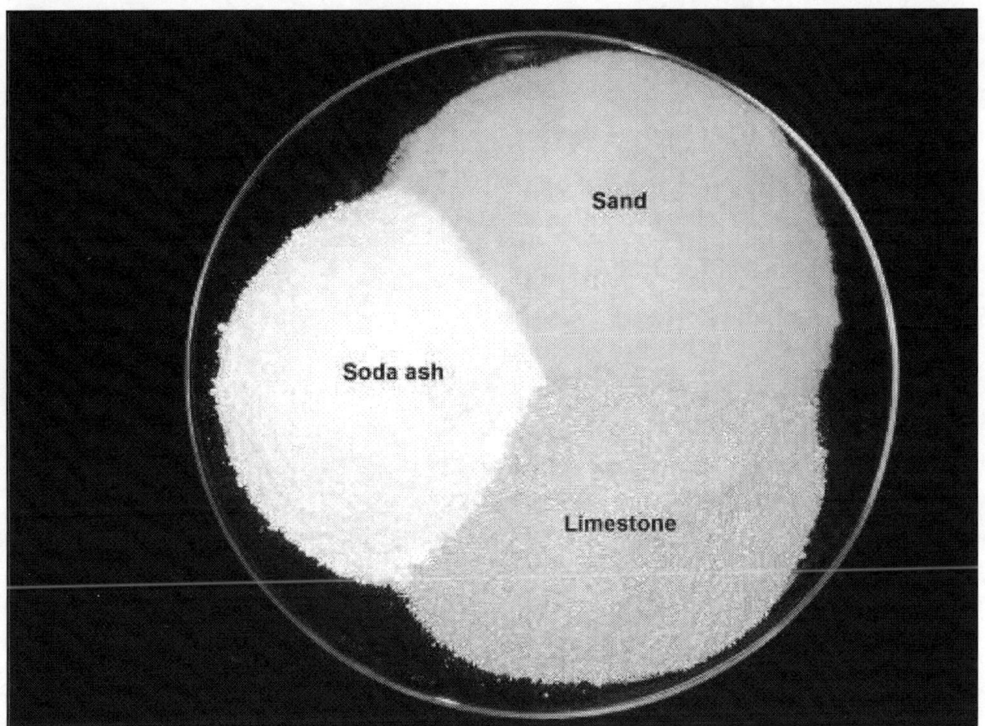

Silica Sand

Three of most common rock forming minerals on earth. Chemically named: quartz sand / rock crystal. Properties: Extremely heat durable and Chemical stack resistance.

Formation of Silica Sand can be natural by mechanical and chemical weathering of quartz-bearing igneous and metamorphic rocks or chemically by weathering of less stable minerals. They break down to become silica sand.

Soda Ash

It is anhydrous sodium carbonate with a texture which is soft and color varying from grey to white. It has an appearance of lump or powder in nature. The formation of Soda Ash can be natural by erosion of igneous rock form sodium deposits, transport by waters as runoffs and collect in basins When sodium comes in contact with CO2, precipitates out sodium carbonate or it can be manufactured by the Solvay process for the manufacture of Soda Ash ($NaHCO_3$).

Lime

This includes hydrated lime Ca $(OH)_2$ and quicklime CaO. Only quicklime CaO can use to make glass.

Manufacture of Glass

The glass container manufacturing process involves a range of procedures, equipment, materials and team members. The key to maintaining optimum control of process quality and productivity is the elimination of any process variation.

Cullet

Recycled glass (from plant and post consumer) used at levels as high as 80 per cent when available. It is needed and added to enhance the melting rate and it significantly reduces energy required for glass production.

Glass Container Recycling

- 100 per cent recyclable. Can be recycled again and again with no loss in quality or purity
- In 2005, 25.3 per cent of glass container recycled
- Good for the environment
- Recycling glass reduces consumption of raw materials, extends the life of plant equipment, and saves energy
- Lighter weight. More than 40 per cent lighter than 20 years ago.

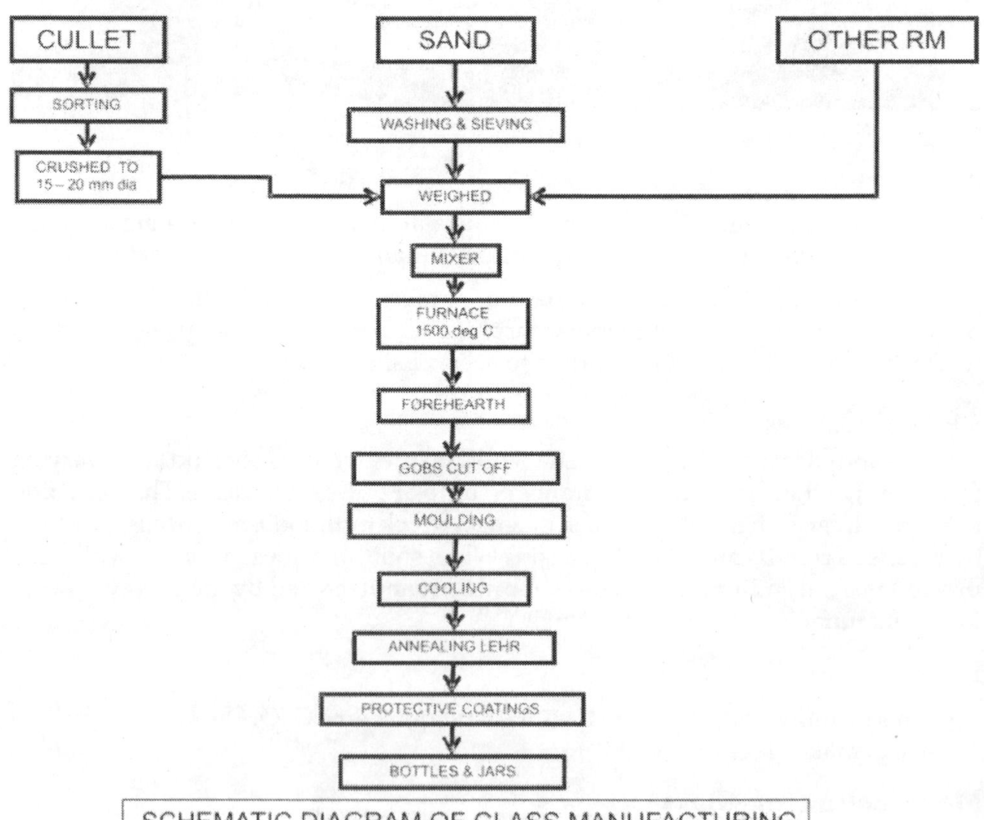

SCHEMATIC DIAGRAM OF GLASS MANUFACTURING

Benefits of Using Quality Cullet

- Over a ton of natural resources are saved for every ton of glass recycled.
- Energy costs drop about 2-3 per cent for every 10 per cent cullet used in the manufacturing process.
- For every six tons of recycled container glass used, one ton of carbon dioxide, a greenhouse gas, is reduced.
- Glass has an unlimited life, it can be recycled over and over again.
- Lesser sodium oxide stronger the glass
- Aluminium oxide increases the hardness and durability.
- Use of Na_2SO_4 and Arsenic reduces blisters

Step 1: Glass Melting

The furnace melts cullet (crushed, recycled glass), sand, soda ash, limestone, and other raw materials together. Molten glass usually ranges in temperature between 2,300 and 2,800° F. A Furnace Control Room houses the computer which monitors and controls furnace temperature.

Cullet + SAND + OTHER RM MELTED in furnace (1500°C) (100 to 500 MT)

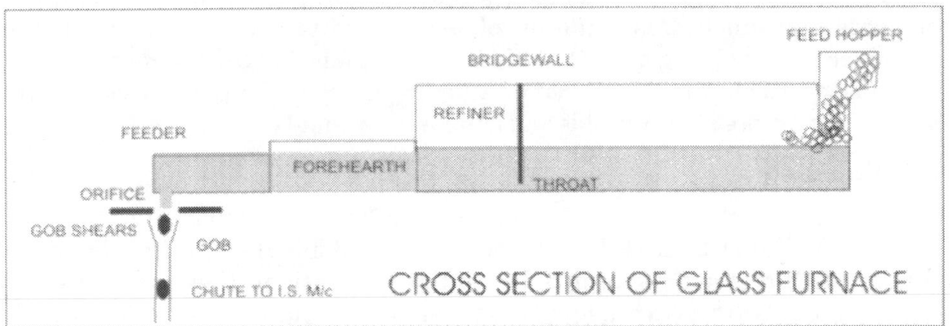

Colour agents added in melt or forehearth
Glass has no distinct melting or solidifying temperature
Decolorizers are added to remove the colour by mineral impurities

Step 2: Container Forming

The Refiner distributes the molten glass to the fore hearth, which brings the temperature of the molten glass to a uniform level. A Shearing and Distribution System cuts molten glass from the fore hearth into uniform gobs and sends them to an I.S. (Individual Section) Forming Machine that forces the molten gobs into the mold shape. The glass temperature drops further in the Forming Machine to below 2,100°F. Formed glass containers leave the machine, crossing a cooling plate where they are cooled rapidly to below 900° F. The glass has now passed from liquid to solid form.

GOB FORMATION

Gobs —to form blank mold

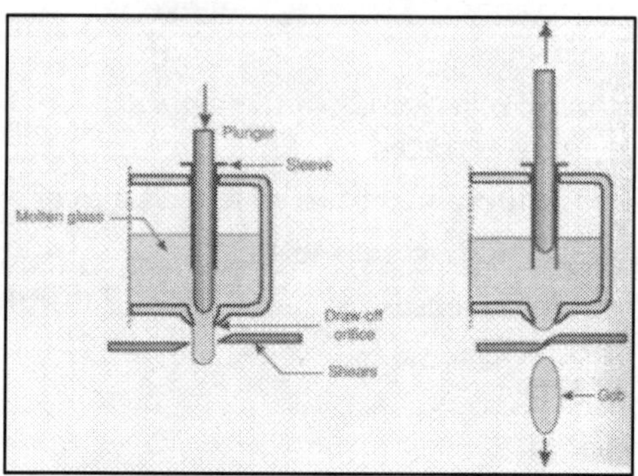

Furnace draw-off orifice and gob shears
Orifice 12 mm to 50mm

Gob is one individual mass of molten glass which makes one container. Molten glass flows depending on the bottle size. Mechanical shears snip off "gobs" of molten glass. Each makes one container. Falling gob is caught by spout and directed to blank molds. Mass-production is made up of several individual sections, each is an independent unit holding a set of bottle-making molds. Large bottles consists of a blank mold and a blow mold. Higher production using double or triple gobs on one machine. two or three blank molds and similar blow molds.

Glass Moulding

BLOWING (Bottle or Jar) is a TWO STAGE MOULDING with BLANK MOULD and BLOW MOULD. BLANK MOLD forms neck and initial shape. Parison mould where gob falls and neck is formed. has number of sections, finish section, cavity section (made in two halves to allow parison removal), a guide or funnel for inserting gob, a seal for gob opening once gob is settled in mold, blowing tubes through the gob and neck openings BLOW MOULD produces the final shape.

There are TWO TYPES OF PROCESSES.

BLOW & BLOW	PRESS & BLOW	

The two processes differ according to the parison producing.

BLOW and BLOW

BLOW and BLOW Process for Narrow-Necked Bottles

- Gob dropped into the blank mold through a funnel-shaped guide (985°C)
- Parison bottomer replaced guide ;air blown into settle mold to force the finish section. At this point the bottle finish is complete.
- Solid bottom plate replaced parison bottomer and air is forced to expand the glass upward and form the parison.
- Parison removed from the blank mold, rotated to a right-side-up orientation for placemcnt into thc blow mold.
- Air forces the glass to conform to the shape of the blow mold. The bottle is cooled to stand without becoming distorted and is then placed on conveyors to the annealing oven.

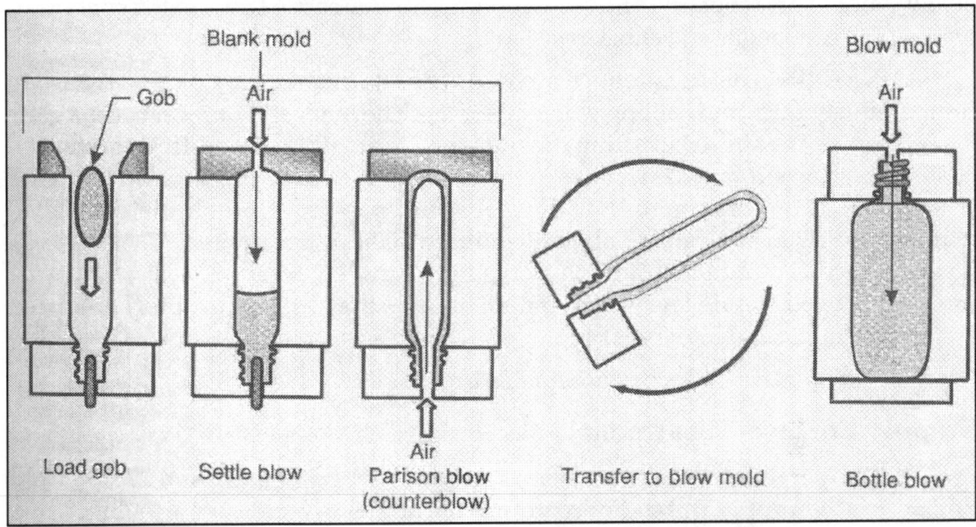

PRESS and BLOW

Press-and-Blow Process for Wide-Mouthed Jars

Gob delivery and settle-blow steps are similar to blow-and-blow forming. Parison is pressed into shape with a metal plunger rather than blown into shape The final blowing step is identical to the blow-and-blow process. Used for smaller necked containers. Better control of glass distribution

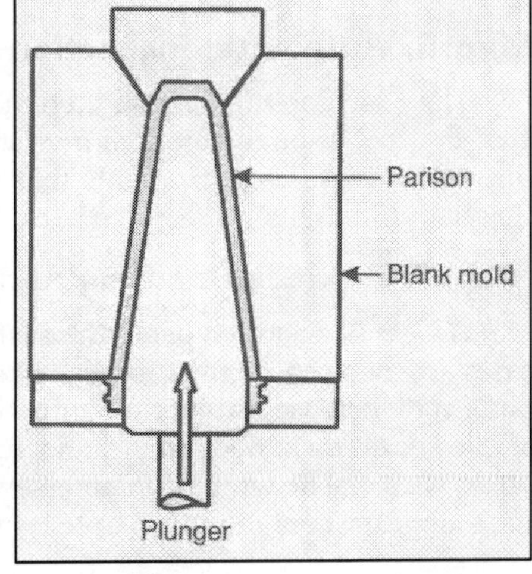

Difference of the Two Processes

Blow-and-blow used for narrow-necked bottles.Press-and-blow used to make wide-mouthed jars and for increasingly smaller necked containers and it has better control of glass distribution. Typical production rates range from 60 to 300 bottles per minute, depending on the number of sections in a machine and the number of gobs being extruded, and the size of the container.

The blown bottle is removed from the blow mold with takeout tongs and placed on a deadplate to air cool for a few moments before transfer to a conveyor that transports it to the annealing oven.

Step 3: Container Conditioning

The formed containers are loaded into an Annealing Lehr, where their temperature is brought back up close to the melting point, then reduced gradually to below 900° F. This reheating and slow cooling eliminates the stress in the containers making them stronger and shock resistant.

ANNEALING helps to reduce internal stresses; in annealing oven. - Walls are comparatively thick and cooling will not be even. The inner and outer skins of a glass become rigid. The still-contracting inner portion build up internal stresses. Uneven cooling develop substantial stresses in the glass. Bottle passes through an lehr after removal from the blow mold. - LEHR is a belt passing through the controlled temperature oven at a rate. of about 200mm to 300mm per minute. Glass temp is raised to 565° C and then gradually cooled to room temperature with all internal stresses reduced to safe levels in about an hour as they exit. Improperly annealed bottles are fragile and high breakage.

Hot-filling also produce unacceptable breakage levels.

Step 4: Surface Treatment

The temperature of the containers is reduced to between 225 and 275° F. Cold End Sprays then apply an exterior coating to the bottles to increase line mobility, and reduce abrasions to maintain the inherent strength of the container

Step 5: Automatic Inspection

The Fast Cooling Section then brings container temperatures down to about 100° F - cool enough to touch by hand. The manufactured containers then pass through a series of instruments that physically and optically test the containers. Rejected containers are recycled back into the furnace.

Step 6: Product Handling and Packaging

A Case Packer then packs the containers in corrugated cases for shipment. The cases are then sent to the Case Palletizer, where they are stacked in a prearranged pattern to increase stability for shipment. A strapper fits plastic bands around the stacked boxes for added stability and, finally, the Stretch Wrap Unit covers the stacked boxes with plastic wrap. Containers can also be sent to a Bulk Palletizer that stacks the individual containers in 5 to 15 layers, depending upon the size of the container.

These Bulk loads are also strapped for stability and sent through the Stretch Wrap Unit and covered with plastic wrap for shipment.

Shape and Strength of Glass Container

The shape and strength of the container are interrelated. The glass container having sharp corners will have the least strength to pressure.

For a given weight of glass, a container having spherical shape will be the strongest followed by elliptical shape and square shape with rounded corners.

Decoration of glass

Glass can be decorated in many ways; it can be screen printed, vacuum metallised, labeled, electroplated and vinyl coated.

Printing labels in permanent colours directly on the glass is done by a process called Applied Ceramic Labelling.

Glass Container Properties

- Is impermeable and nonporous. Safeguards against moisture and oxygen invasion. Sanitary and odorless.
- Can be irradiated–an attributed for sterilized applications for the future.
- Is recyclable. Does not pollute. Important environmental consideration.
- Combined with foil, can make an effective tamper-evident container.
- Is ideal for high-speed filling lines. Speed means more profit.
- Takes high-torque sealing. Helps safeguard against leakers in shipment.
- Is resealable. Unused products can be saved for future use.
- Does not deteriorate, corrode, stain or fade.
- Is microwavable–(depending on the specific type of jar).
- Needs no protective coating inside. Products stay uncontaminated.
- Is stackable on shelves. Makes a neat and uniformly good impression.
- Is economical, when advantages are weighed against other packages.
- Is retortable. Can be used to process foods and other products at high temperatures.
- Is a proven package for strong chemicals and solvents.
- Is transparent. It can showcase a product and entice the buyer.
- Has FDA acceptance. Considered a safe packaging material.
- Is made from abundant raw materials. A reliable supply is assured.
- Is available in stock or private molds. Stock molds can be economically customized in appearance.
- Can achieve a hermetic seal and prolong product shelf life.
- Creatively shaped or decorated. can add consumer value after product use–specifically for crafts or collectibles.

- In amber or green, can filter out harmful ultraviolet rays. Protects product integrity.
- Is attractive. Connotes richness and substance. Provides a quality image for products.
- Has proven customer appeal. confirmed by independent tests.
- Is compatible with child-resistant closures.
- Is autoclavable. Great for injectible and other pharmaceutical products.
- Can be labeled, colored, silk screened, enameled, etched, sand blasted or coated. Decoration opportunities abound.
- Has almost a zero rate of transmission. Guards against loss of moisture, product strength, freshness and aroma.
- Can be formed into various sizes and shapes. Can provide a strong family identity for product lines.

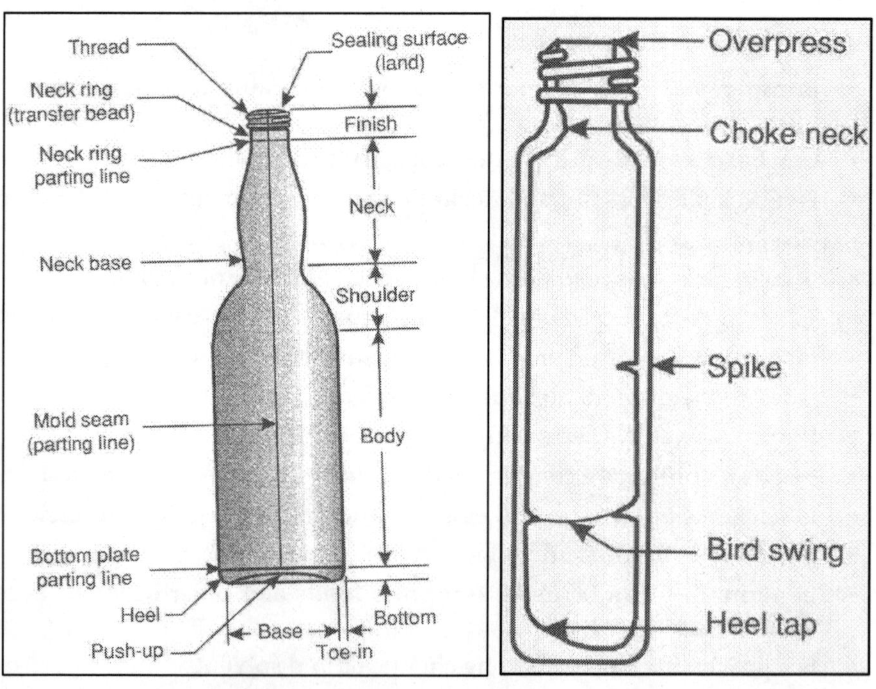

Bottle Parts **Bottle Defects**

Types of Glass

Although INERT Sodium and other ions can leach out on ceratin solution.

USP Type-I Borosilicate Flint (clear), Amber (brown) glass vials, ADDITION OF 6 per cent BORON REDUCES LEACHING ACTION. Least reactive glass available for containers. It can be used for all applications and is most commonly used to packaged water for injection, UN-buffered products, Chemicals, sensitive lab samples, and samples requiring sterilization. All lab glass apparatus is generally Type I borosilicate glass. Type I glass is used to package products which are alkaline or will become alkaline prior to their expiration date.

USP Type-II De Alkalized Soda Lime Glass(type3) that has been treated in the lehr with sulphur to reduce alkali solubility. The treatment produces a disccoloured appearance. Has higher levels of sodium hydroxide and calcium oxide.It is less resistant to leaching than Type I but more resistant than Type III. GOOD ALKALI RESISTANCE. It can be used for products that remain below pH 7 for their shelf life

USP Type-III conventional soda glass. Acceptable in packaging some dry powders which are subsequently dissolved to make solutions or buffers. It is also suitable for packaging liquid formulations that prove to be insensitive to alkali. Type III glass should not be used for products that are to be autoclaved, but can be used in dry heat sterilization

USP TYPE NP SODA LIME GLASS. Is a general purpose glass and is used for non-parenteral applications where chemical durability and heat shock are not factors. These containers are frequently used for capsules, tablets and topical products.

Performance and Testing

It is important that containers comply with specification and general industry guidelines in order to withstand the normal stresses and mechanical abuse right through until the end user has finished using it.

Vertical Load

Forces of this nature might be produced during capping or through stacking products on top of each other. To help ensure glass containers have adequate vertical load strength, we test to BS EN ISO 8113-2004 using a Universal Testing Machine.

Impact Testing

To help ensure glass containers have an adequate impact resistance, we can test to standard manufacturing codes of practice using an industry standard Pendulum Impact Tester.

Thermal Shock

Hot-fill or heat-treated glassware can be tested for thermal shock resistance to ensure the product is fit for the intended purpose. Testing can be carried out to ASTM C149 and BS EN ISO 7459 either as pass/fail test typically at 42OC downshock or progressive testing to complete sample failure. Effect of sudden temperature change effect is minimal if both sides are heated or cooled simultaneously effect is prominent when one surface is hot and the other surface is chilled

Coating Performance

Assessment of surface protection can be carried out by use of slip tables and hot end coating technology. The longevity of the coating performance can be assessed using line simulator, whereby bottle to bottle abrasion damage which may be expected to occur on a filling line can be replicated and the subsequent damage of the container tested. This is of particular use for returnable glassware.

Internal Pressure Resistance

Carbonated beverage bottles need to be able to withstand without failure the pressure produced by their contents over long periods.

Residual Strain

Measurement of annealing stresses / residual strain to ASTM C148.

Online Inspection of Glass Bottles

1. Bottle Spacer. This machine is pre-set to create a space between the bottles on the conveyer to avoid bottle to bottle contract.
2. Squeeze Tester. Each bottle is passed between discs that exert a force to the body of the container. Any obvious weakness or crack in the bottle will cause it to fail completely with the resulting cullet being collected by a return conveyor running underneath.
3. Bore Gauger. The internal and external diameter at the neck finish entrance to the bottle and the bottle height are measured. Bottles outside specification are automatically rejected by means of a pusher positioned downstream from the gauger.
4. Check Detector. Focuses a beam of light onto areas of the container where defects are known to occur from previous visual examinations, any crack will reflect the light to a detector, which will trigger a mechanism to reject the bottle.
5. Wall Thickness Detector. This test uses dielectric properties of the glass, the wall thickness can be determined by means of a sensitive head which traverses the body section of the container. A trace of the wall thickness is then obtained and bottles falling below a specified minimum will be automatically rejected.
6. Hydraulic Pressure Tester. A test carried out on bottles which will be filled with carbonated beverages and gauges the internal pressure of every bottle before it is packed.
7. Visual Check. Bottles are passed in front of a viewing screen as a final inspection.
8. Glass failure is usually as a result of thermal shock or impact stresses. Each glass container has a maximum thermal expansion threshold and a maximum vertical load stress, which it can withstand without cracking. These values should be known before it is used for a particular application.
9. The shape of the container will influence its strength, smooth edges result in the formation of a stronger container than one with rectangular or sharp edges.

8
Regenerated Cellulose Films

History

1905

Jacques Brandenberger, a Swiss chemist, was not trying to make something to cover your pork chops in 1908. He worked in a French textile firm and was looking for a way to make a stain proof tablecloth. He tried coating the cloth with a thin sheet of viscose film. No one would buy the tablecloths, but Brandenberger realized that the sheet of film held possibilities.

1913 Commercial Production

It took him ten years to develop a machine that would produce what he named cellophane. He patented the production process and called the company La

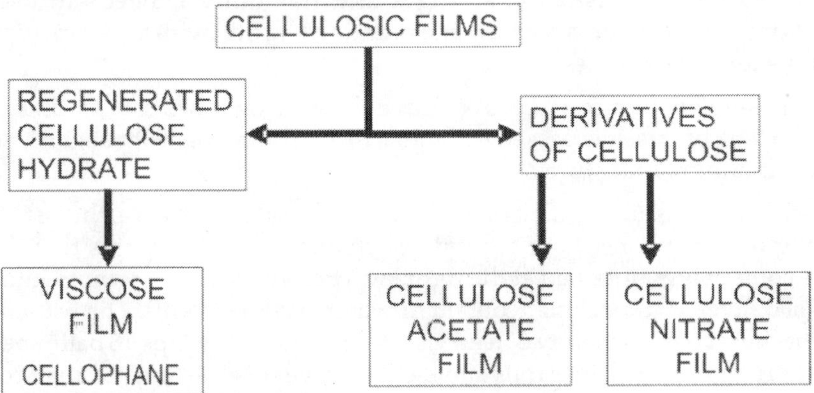

Cellophane. The name came from combining "CELLO," from cellulose, with "PHANE," from the French word diaphane, which means transparent. Cellophane became available to the public in 1919.

1927 Waterproofing

In 1927 W.H.Charch and W.E.PindLe (USA) developed a waterproof lacquer coating that made it more useful. With the lacquer coating, cellophane could be used to package food, since it was airtight and waterproof.

Manufacture in a Nutshell

In the manufacturing process, an alkaline solution of cellulose fibers (usually wood or cotton) known as viscose is extruded through a narrow slit into an acid bath. The acid regenerates the cellulose, forming a film.

Further treatment, such as washing and bleaching, yields cellophane. Between the two, viscose is trans-formed into cellulose film. In the passes, it passes 266 special designed rollers and through a series of baths which either treat or clean the film.

At the casting lips the first property of the finished film its thickness variation is controlled by the use of adjuster bolts which vary the distance between the lips and the thickness itself is controlled by varying the rate at which the viscose is pumped through by the metering pump relative to machine speed. When viscose is extruded into a bath of sulphuric acid it forms a continuous wed of re-generated cellulose film. At this stage the film is very pliable and flimsy, and is quite opaque.

A number of by-products formed as the film passes through the acid bath are then removed by washing the film with almost boiling water, and passing it through various chemical treatment baths including a bleach. Carbon disulphide released in this process is recovered.

After the bleach bath the wed is washed thoroughly with a counter current flow of water.The continuous wed has become pure cellulose film. However in this form it is too brittle and inelastic for most of the end uses normally associated with Cellophane.

Therefore it is further treated by being passed through softening baths containing chemical agents such as glycerin or glycols.

This treatment gives it the strength and flexibility it needs in use, and modifications of the softeners can be made to endow the film with the properties most appropriate for different uses.

By this time the film contains over three times its own weight of water and, to remove the bulk of this, it is passed through a dryer containing 90heated rollers and removed superfluous moisture.

Because the moisture content of the film is of critical importance, the last stage of the casting machine consists of a humidity chamber which ensures that the film moisture content is maintained at the right level before the film is wound into a mill roll,and is delivered to packing.(content of water is 7-8 per cent). The wet-end of a casting machine consists mainly of feed viscose system,casting lips,18 bath, specially-designed driver or mot-driver rolls and etc. The dryer of a casting machine consists mainly of housing, 90 heated rolls, 84 heat-air engine, humidity chamber,exhaust of heated rolls,insulated cover, slitter, wind up system and etc.

A. RAW MATERIALS

- HIGH PURITY WOOD PULP
- HIGH PURITY CAUSTIC SODA
- SULPHURIC ACID
- CARBON-DI-SULPHIDE
- SODIUM SULPHATE
- SODIUM SULPHIDE
- BLEACH LIQUOR
- ANTICHLOR MATERIAL
- GLYCERINE and PLASTICERS
- SURFACE ACTIVE AGENTS
- SURFACE COATING MATERIALS
- SOLVENTS
- DYES and PIGMENTS

B. PROCESS

- STEEPING
- SHREDDING and MATURING
- XANTHATING and DISSOLVING
- RIPENING, FILERATION and DE-AREATION
- FILM CASTING
- STEEPING
    - PULP SHEETS STEEPED IN 20 per cent NaOH
    - SWELLS and FORMS ALKALI CELLULOSE
    - PRESSED TO REMOVE EXCESS ALKALI
- SHREDDING and MATURING
    - ALKALI CELLULOSE SHEETS
    - SHREDDING MACHINES
    - SHEET BROKEN TO FLUFFY POWDER
    - ALKALI CELLULOSE CRUMB (150/200g/Lt)
    - FED TO SLOW ROTATING GIANT DRUM
    - AGING/MATURING FOR 12 – 24 hrs (OXYGEN ABSORBED, CELLULOSE CHAIN LENGTH SHORTENED)
- XANTHATING and DISSOLVING
    - ALKALI CELLULOSE CRUMB

  - SPIRALLY DESIGNED ROTATAING CHURNS
  - ORANGE COLOURED MASS – CELL. XANTHATE
  - DISSOLVE IN NaOH (20 per cent) IN 3-5 hrs
  - HONEY COLOURED LIQUID
  - VISCOSE 7-8 per cent, 5-6 per cent NaOH
    R OH + NaOH = R O Na + H2O (R CELLULOSE)
    R O Na + CS2 = RO-CS-SNa
- RIPENING
  - RIPENED 20 – 40 hrs
  - LOW TEMPERATURE
  - FILTERED
  - FOR COAGULATION and VISCOSITY
  - FOR CASTING
- CASTING
  - RIPENED VISCOSE SOLUTION
  - PUMPED
  - HOPPER
  - TIP(100-200-400microns) IMMERSED IN H_2SO_4 – NaOH AT 35-40 deg
  - VISCOSE MASS FORCED THROUGH COAGULATION AND FORMS SHEET FILM
- WASHED
  - WASHING (H2O) DESULPHURING, ALKALI,BLEACHING.
  - WASHING AND SOFTENING WITH GLYCERINE
  - DRIED, HUMIDITY CONDITIONED AND WOUND
  - PT FILM ————————— COATING

Nomenclature

CODE	DESCRIPTION
A	ANCHORED
C	COLOURED
D	DEMI COATED (ONE SIDE)
H	RESISTANT TO BLOCKING IN HUMIDITY
J	FLAME RESISTANT
L	LESS MOISTUREPROOF THAN STANDARD
M	MOISTUREPROOF
O	OPAQUE
P	PLAIN

R	RANCIDITY RETARDANT
S	HEAT SEALABLE
T	TRANSPARENT
U	UNSIZED, LOW SURFACE SLIPPAGE
XX	POLYMER COATED

GRAMS PER 10 SQM

- NORMAL 300,400,500 **(INDICATES WEIGHT PER 10 sqm)**

TYPE	**GSM**
300 PLAIN	29-34
300 COATED	32-37
400 PLAIN	38-45
400 COATED	40-47
600 PLAIN	56-64
600 COATED	60-68

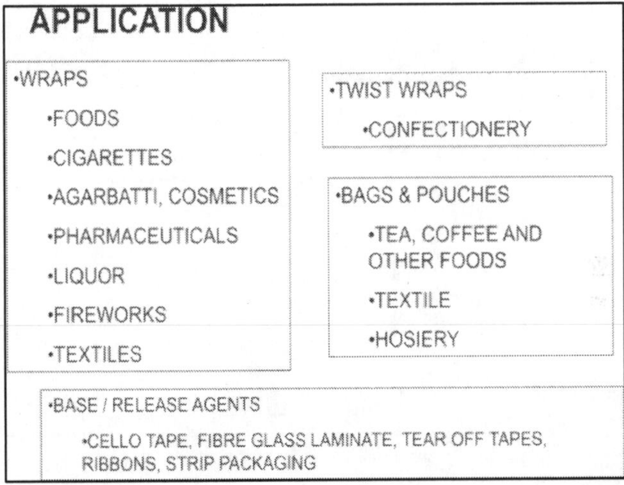

CONVERSION

- METHODS AND FORMS
  - COATING
  - LAMINATION
  - PRINTING
  - CONVERSION TO BAGS, POUCHES, TAPES
- COATING
  - WAX and HOT MELT

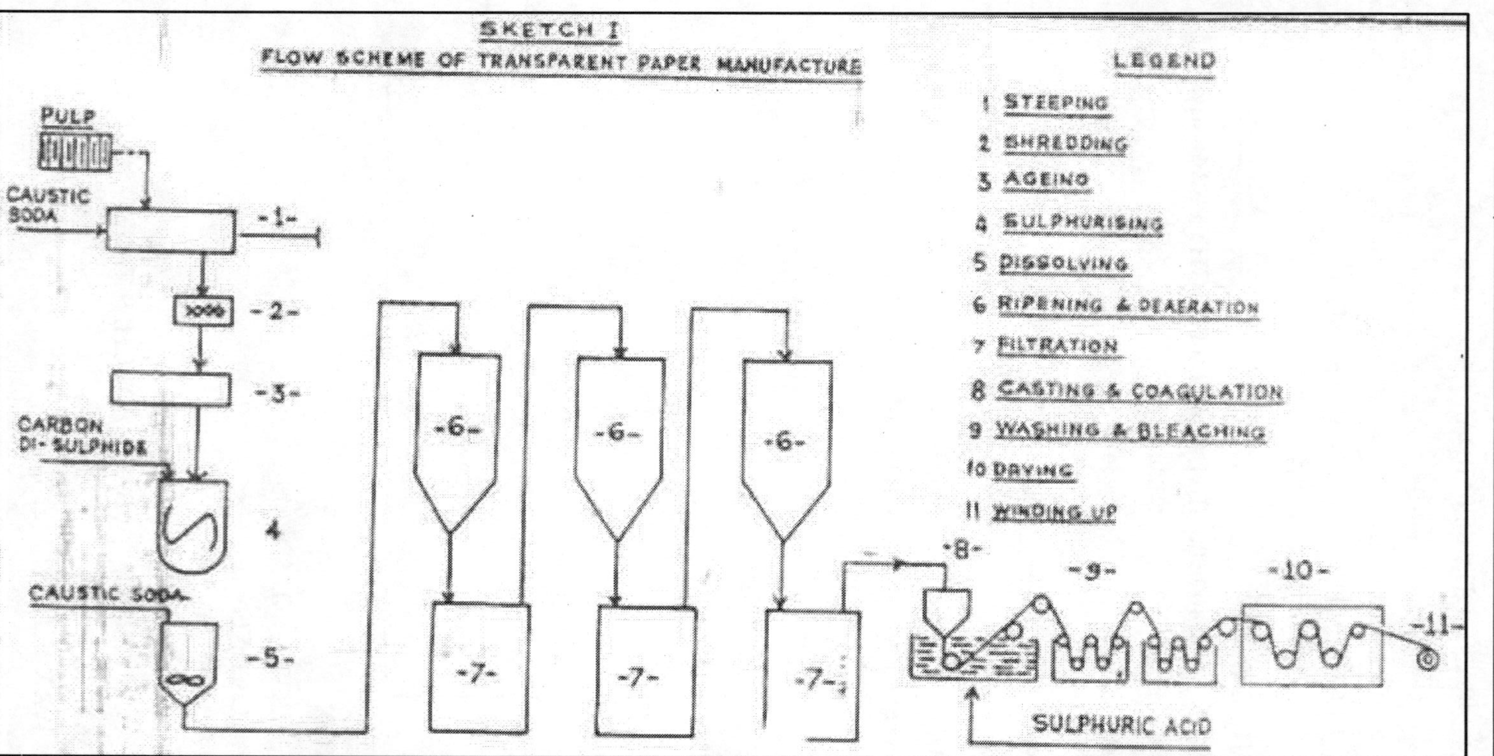

Figure 12.3: Schematic Diagram of the Modied BSF Device.

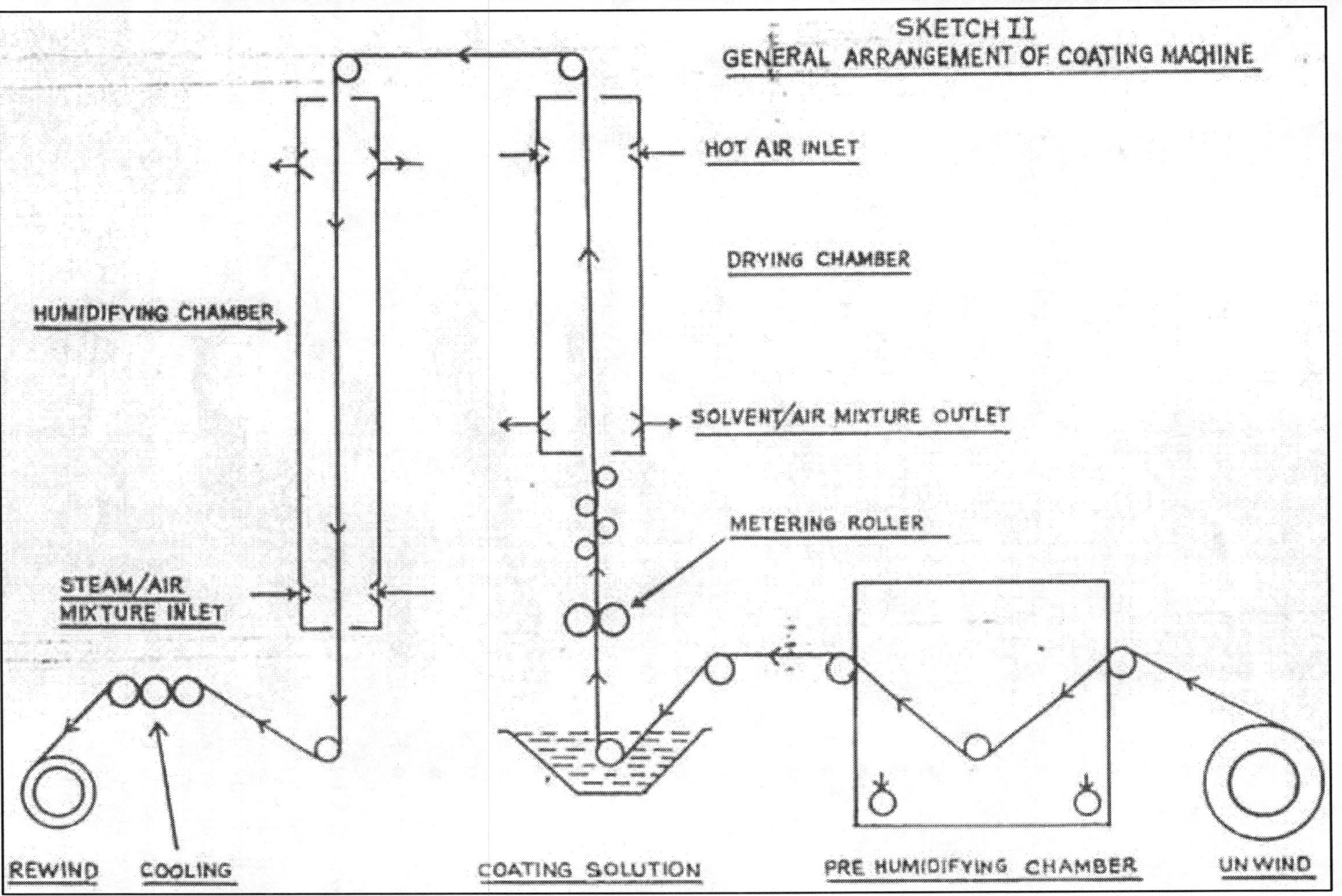
SKETCH II
GENERAL ARRANGEMENT OF COATING MACHINE
HOT AIR INLET
DRYING CHAMBER
HUMIDIFYING CHAMBER
SOLVENT/AIR MIXTURE OUTLET
METERING ROLLER
STEAM/AIR
MIXTURE INLET
REWIND
COOLING
COATING SOLUTION
PRE HUMIDIFYING CHAMBER
UN WIND

  - DISPERSION COATING
  - NITRO CELLULOSE LAQUER
  - POLYMERS / CO-POLYMERS
- LAMINATION
  - DRY, HEAT, WAX and WET LAMINATION
- PRINTING
  - FLEXOGRAPHY AND GRAVURE
- METALLISNG
  - WITH ALUMINIUM FOIL and HIGH VACUUM PROCESS
- STORAGE - HANDLING
  - 35-45 per cent RH, 18-24 deg C
  - AVOID DIRECT EXPOSURE TO SUNLIGHT
  - AVOID EXTREME TEMPERATURES
  - FILM ROLLS NOT TO BE DROPPED
  - ALWAYS STORE IN MOISTUR BARRIER MATERIALS
  - ALWAYS STORE BY THE CORE AT EACH END
- PROPERTIES
  - SPARKLING TRANSPARENCY
  - UNIFORM QUALITY
  - EXCELLENT MACHINEABILITY
  - AMENABLE TO PRINTING and CONVERSION
  - GOOD BARRIER9DEPENDS ON TYPE)
  - GOOD OIL / GRESEPROOFNESS
  - SHRINKAGE 3-5 per cent
  - NORMAL SEAL TEMPERATURE 100 – 130°C
  - TEAR STRENGTH – LOW
  - FLEXIBILITY - GOOD

Jute, Cotton and Woven Fabric

Jute is a long, soft, shiny vegetable fibre that can be spun into coarse, strong threads. It is produced from plants in the genus Corchorus, which has been classified in the family Tiliaceae, or more recently in Malvaceae. Jute is one of the most affordable natural fibres and is second only to cotton in amount produced and variety of uses of vegetable fibres. Jute fibres are composed primarily of the plant materialscellulose (major component of plant fibre) and lignin (major components of wood fibre). It is thus a ligno-cellulosic fibre that is partially a textile fibre and partially wood. It falls into the bast fibrecategory (fibre collected from bast or skin of the plant) along with kenaf, industrial hemp, flax(linen), ramie, etc. The industrial term for jute fibre is *raw jute*. The fibres are off-white to brown, and 1–4 metres (3–12 feet) long.

Jute needs a plain alluvial soil and standing water. The suitable climate for growing jute (warm and wet climate) is offered by the monsoon climate during the monsoon season. Temperatures from 20C to 40C and relative humidity of 70 per cent –80 per cent are favourable for successful cultivation. Jute requires 5–8 cm of rainfall weekly and more during the sowing period.

For centuries, jute has been an integral part of culture of Bengal, in the entire southwest of Bangladesh and some portions of West Bengal. During the British Raj in the 19th and early 20th centuries, much of the raw jute fibre of Bengal was carried off to the United Kingdom, where it was then processed in mills concentrated in Dundee. Initially, due to its texture, it could only be processed by hand until it was discovered in that city that treating it with whale oil, it could be treated by machine. The industry boomed ("jute weaver" was a recognised trade occupation in the 1901 UK census), but this trade had largely ceased by about 1970 due to the appearance of synthetic fibres.

Margaret Donnelly, a jute mill landowner in Dundee in the 1800s, set up the first jute mills in Bengal. In the 1950s and 1960s, whennylon and polythene were rarely used, one of the primary sources of foreign exchange earnings for the erstwhile United Pakistan was the export of jute products, based on jute grown in then East Bengal now Bangladesh. Jute has been called the "Golden Fibre of Bangladesh." However, as the use of polythene and other synthetic materials as a substitute for jute increasingly captured the market, the jute industry in general experienced a decline.

During some years in the 1980s, farmers in Bangladesh burnt their jute crops when an adequate price could not be obtained.

Many jute exporters diversified away from jute to other commodities. Jute-related organisations and government bodies were also forced to close, change or downsize. The long decline in demand forced the largest jute mill in the world (Adamjee Jute Mills) to close in Bangladesh. Bangladesh's second largest mill, Latif Bawany Jute Mills, formerly owned by businessman, Yahya Bawany, was nationalized by the government. Farmers in Bangladesh have not completely ceased growing jute, however, mainly due to demand in the internal market. Between 2004–2010, the jute market recovered and the price of raw jute increased more than 500 per cent [citation needed

Jute has entered many diverse sectors of industry, where natural fibres are gradually becoming better substitutes. Among these industries are paper, celluloid products (films), non-woven textiles, composites (pseudo-wood), and geotextiles.

In December 2006 the General Assembly of the United Nations proclaimed 2009 to be the International Year of Natural Fibres, so as to raise the profile of jute and other natural fibres.

Jute is a rain-fed crop with little need for fertilizer or pesticides. The production is concentrated in [Bangladesh] and some in India, mainly Bengal. The jute fibre comes from the stem and ribbon (outer skin) of the jute plant. The fibres are first extracted by retting. The retting process consists of bundling jute stems together and immersing them in low, running water.

There are two types of retting: stem and ribbon. After the retting process, stripping begins. Women and children usually do this job. In the stripping process, non-fibrous matter is scraped off, then the workers dig in and grab the fibres from within the jute stem. India, Pakistan, China are the large buyers of local jute while Britain, Spain,Ivory Coast, Germany and Brazil also import raw jute from Bangladesh. India is the world's largest jute growing country.

Jute Bags

Traditionally, jute has been the packaging material used for bulk packaging. With the increasing demand for bulk Packaging and due to the stagnant jute production, plastic woven sacks have the potential to fulfill this need in a cost-effective manner. These are made either from HDPE or PP.

Disadvantages of Jute Bags

- Availability: Food grade jute bags are not easily available.
- Mineral Oil Contamination: Mineral oil is used as a lubricant in processing of jute fibres. Mineral oil adds hydrocarbon odour to the fabric and some free radicals are also found in mineral oil which can affect the product packed.
- Poor resistance to UV radiation.
- Susceptibility of the material to insect infestation and rotting.
- Poor resistance to corrosive chemicals.
- Insect Breeding: It is a regular practice in the grain trade to reuse bags. The structure of jute fabric being porous, the insects like stored grains and pests find it easier to lay their eggs on its fibres. The eggs of stored grain pests are microscopic in nature and one can know about the presence of infestation only when it gets developed into the larva stage. Such bags can therefore serve as a potential source of infestation.
- Cost: Jute sacks are 5-6 times heavier than the sack made out of plastic material like HDPE or Polypropylene for a similar weight pack. The jute goods are transported all the way from east India to other parts of the country, which are thousands of miles away. Owing to these factors, the cost of jute bags is much higher than the plastic sacks.

The nature of jute packaging is such that, lot of food packed therein gets exposed to deteriorating factors and germs. Air borne germs and the ones present in the storage godowns may seek way through the pores of the fabric and may contaminate the food. Such food, when consumed may cause illness like food poisoning. High Density Polyethylene (HDPE) / Polypropylene (PP) Woven Sacks

HDPE and PP woven sacks have replaced jute bags in number of applications. Several plants manufacturing these sacks have come up in different parts of the country making the availability of these products possible at low price.

Advantages of HDPE/PP Woven Sack

- Elongation at break of HDPE tapes is about 15-20 per cent in comparison to jute bags, which is about 30 per cent. Owing to this property HDPE woven sacks have better resistance to dropping.
- HDPE / PP woven sacks do not impart any odour to the food product packed in them
- HDPE / PP woven sacks are not attacked by insects
- HDPE and PP woven sacks of strength equivalent to that of jute bags can be made using almost 70 times lower weight of the resin and hence are almost 60-65 per cent cheaper than the jute bags.
- HDPE and PP woven sacks are the most hygienic material for packing of cereals and pulses and one need not reuse the same owing to their low cost.
- Fabric allows diffusion of air / gases easily through the gaps between the filament thus facilitates ventilation of grains during earlier stages of harvest and penetration of fumigants.
- Although HDPE / PP undergo degradation under UV light, it is possible to arrest the same by using appropriate UV stabilizers.
- It is possible to laminate HDPE woven sacks with LDPE. The laminated bags protect the product packed in the bag from moisture and also prevent the loss of products like flours due to spillage, which usually occur through plain jute bags, which are commonly used for packing of flour.

HDPE / PP Woven Sacks Lamination of PP woven sacks with polypropylene is also possible, as the grade of PP suitable for lamination is now available in the market.

Textiles

Textile containers have poor gas and moisture barrier properties and have a poorer appearance than plastics. Woven jute sacks, which are chemically treated to prevent rotting and to reduce their flammability, are non-slip, have a high tear resistance, and good durability. They are used to transport a wide variety of bulk foods including grain, flour, sugar and salt.

Cotton

Calico is usually a closely woven, strong, plain, cotton fabric which is inexpensive and is satisfactory as a wrapper for flour, grains, legumes, coffee beans and powdered or granulated sugar. It can be re-used as many times as the material withstands washing and is easily marked to indicate the contents of the bag.

Muslin and cheesecloth are open-mesh, light fabrics used to wrap soft foods, which they help hold together in the desired shape. Processed meats, smoked shoulders of ham, etc, are tightly wrapped in cheesecloth, before being packaged into cellophane, wax paper etc. Both muslin bags and cheesecloth wrappers have to be cut open and can seldom be re-used. It is a very cheap material, made in huge quantities because of its multiple applications, but it gives very little protection to food, and simply holds it together.

9
Distribution Hazards

Distribution

Goods are produced to be consumed. The major portion of the price we pay to a product is for the distribution. This portion is as high as 50 per cent of the price of the product. Distribution bridges the gap between the manufacture and consumption. Distribution activity comprises of *Storage, Handling and Transportation*. These activities by virtue have many dangers or hazards in built in them. In order to take care of these hazards, Packaging acts as a coordinated system of preparing goods for safe, secure, efficient and effective handling, transport, distribution, storage, retailing, consumption

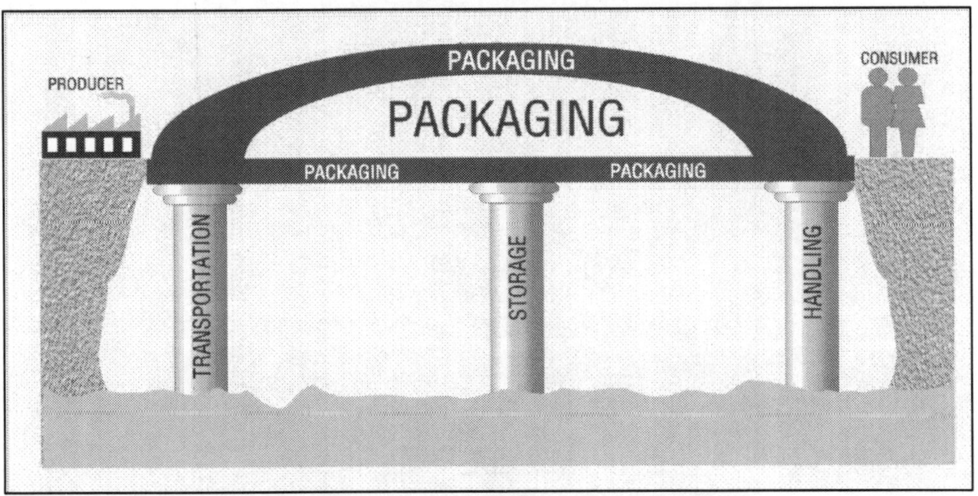

and recovery, reuse or disposal combined with maximizing consumer value, sales and hence profit.

Above it's fundamental function of containing, protecting, preserving, identifying,educating and dispensing the product, the functions of packaging are manifold and complex and the definition here can be related to three main categories *i.e.* logistics, marketing and environment.

Logistical function	Facilitate distribution
	Protect both product and the environment
	Provide information about conditions and locations
Marketing function	Graphic design, format
	Legislative demands and marketing
	Customer requirements/consumer convenience for end use as well as distribution
Environmental function/aspect	Recovery/Recycling
	Dematerialisation
	One-way vs. reusable package
	Toxicity

Packaging may be classified as primary, secondary or tertiary, reflecting the levels of packaging. These definitions should be used together with the consideration of packaging as a system, with hierarchical levels. See Figure below. This approach

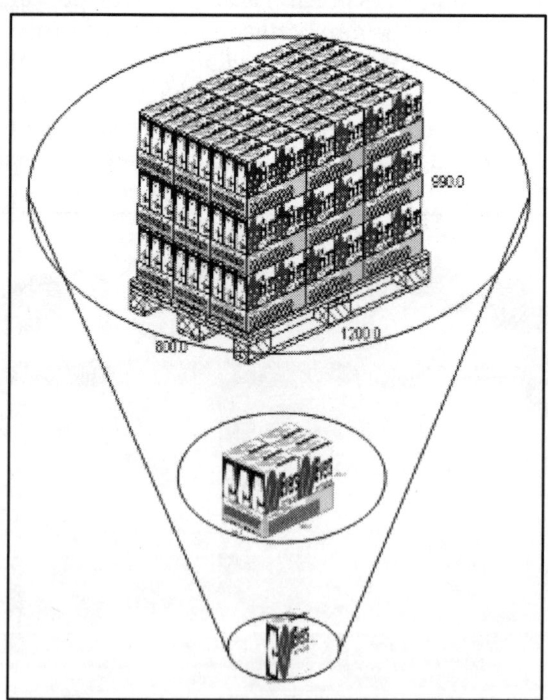

highlights the natural interaction between the different levels of packaging and facilitates an understanding of their interdependence.

Function of Distribution

- Reach product to consumer in factory fresh condition
- Production and consumption are dependent on distribution
- There is always a time lag between production and consumption

Stages of Distribution

Inherently involves certain stages

- Storage
- Handling
- Transportation

These are pillers of bridge of distribution

There is always a time lag during these stages

Process of Distribution

- Hazards exist during the various stages of distribution
- Inherent by virtue of its process
- Hazards exists during the time lag
- Time lag itself is an hazard

Packaging and Distribution

Packaging is a function which attempts to protect the product from hazards ofdistribution. Hence packaging shouldbe aware of the distribution mechanism to take care of the hazards

Where are the Hazards

- Handling
- Transportation
- Storage
- Climatic
- Environmental

Handling Hazards

Vertical Impact

Handling is not limited to the package itself. At the macro level, an entire shipping container may be handled by crane. At the intermediate level, a package may be part of a pallet load. These pallets are usually handled by automated or semi automated equipment. Individual packages are most often handled by hand. Impact or Shock is measured as G force, with 1 G representing the normal force of gravity. The greatest shock, in most cases, occurs during individual package handling.

This hazard of distribution can occur when packages are dropped; strike (or are struck by) other packages or by sorting mechanisms; or shift and fall during transit. In this environment, there are many shocks of relatively low severity (equivalent to drops from relatively low heights), with typically only a few impacts equivalent to drops from 30 to 40 inches or higher.

Although it is recognized that impacts predominately occur on and around the base of the package (bottom face, edges and corners), packages should be designed to protect against impacts from any direction. Base is generally defined as the bottom surface when the package is in its most stable orientation; sorting operations cannot always honor up arrows or orientation labels.

Long, slender packages may be subjected to bridging during sorting. Bridging can occur during conveyor or chute movement when a long package gets supported only by the ends. It may then be susceptible to damage from even relatively mild impacts near its center. Large, flat packages (mirrors, panels) can be more prone than other types of packages to damage from impacts on their faces.

Typically, only corrugated boxes with weights less than 18 to 20Kg travel through the carriers automatic sorting systems. Large, heavy, non-corrugated and non-rectangular packages (or products with *no* packaging, such as tires or spools of wire) are handled in various ways, depending on the carrier and location, and therefore may be subjected to more severe environments.

Transportation Hazards

- Vibration
- Vertical Impact
- Compression
- Horizontal Impact
- Roll

Vibration

Most vibration occurs during transport. Vibration may cause a package's contents to settle, which makes them more vulnerable to shock and compression because of increased void space. Vibration may also cause migration of some types of packing materials, such as expanded polystyrene. Such migration may leave the product resting in the bottom of the box, relatively unprotected from shock. Potentially damaging vibration may occur during transit by road, rail or air. The severity of motion is low compared to shock, but long exposure times (hours or days) can cause abrasion and scuffing, loosening of closures and fasteners, fatigue and cumulative damage. Distribution companies use a number of different transport modes, including trucks of various types, trailers on flatbed railcars, aircraft and others. The modes can have unique vibration characteristics, and packaged products can react differently to each. Long-haul vehicles are loaded to utilize essentially all of the interior space. This means that package orientation within the load cannot be predetermined, and therefore, packages must be able to withstand vibration in any and all directions. Long and flat packages may not be fully supported along their lengths or faces, potentially causing increased stresses.

Storage Hazards

Compression

In-the-box packaging does two different things: One, it protects the product, but two, it maintains the integrity of the box. Compression is the greatest hazard to box integrity, and void fill minimizes this risk. Research shows that packers often use more void-fill material than necessary, just to be safe. Warehouse storage is generally considered in distribution, but packages are stacked in the long-haul transport vehicles and the resultant compression forces are dynamic (varying due to vibration) during transit. Dynamic compression can be several times the magnitude of static compression. Since package orientation within the load cannot be predetermined, packages must be able to withstand this compression in any direction. The average density of loads in long-haul vehicles can range to 12 or more pounds per cubic foot. Since any particular package can be at the bottom of the stack, the usual approach is to design for dynamic forces equivalent to average load densities at full vehicle heights. The carriers local pick-up and delivery vans generally have interior shelves to hold the packages, so the compressive forces in these vehicles are much less. But the vibration signatures are different from the signatures of long-haul, so the potential for damage still exists.

Climatic Hazards

Next to temperature and humidity, variation in atmospheric pressure, particularly during air shipment, is the most common atmospheric hazard. The climatic conditions can affect the characteristics of both products and packages.

High humidity weakens corrugated; high temperatures can weaken plastic, and low temperatures can tend to make it brittle; high altitudes can distort or damage air-containing or sealed components; static discharges can destroy electronics; and so on. Packages must be designed to protect their products under atmospheric extremes, including winter temperatures, summer temperatures plus the temperature rises in closed vehicles and containers, near 100 per cent relative humidity, altitudes to 20,000 feet for unpressurized feeder aircraft and other conditions as appropriate.

- TEMPERATURE (HIGH and LOW)
- HUMIDITY (HIGH and LOW)
- CYCLIC CHANGE OF TEMPERATURE
- CYCLIC CHANGE OF HUMIDITY
- CONDENSATION
- SALT
- RAIN

Environmental Hazards

- MICROBIAL
- INFESTATION
- LIGHT (HIGH INTENSITY)
- SUNLIGHT

Concept of Impacts

Concept of Manual Handling

- LOAD v/s HANDLING and IMPACT
- NORMS IN DISTRIBUTION

Need for Mechanical Handling

- DEVICES
- ULDs
- SYSTEMS CONCEPT
- CONTAINERISATION

10

Packaging Quality Control

The quality movement can trace its roots back to medieval Europe, where craftsmen began organizing into unions called guilds in the late 13th century. Until the early 19th century, manufacturing in the industrialized world tended to follow this craftsmanship model. The factory system, with its emphasis on product inspection, started in Great Britain in the mid-1750s and grew into the Industrial Revolution in the early 1800s.

In the early 20th century, manufacturers began to include quality processes in quality practices.

After the United States entered World War II, quality became a critical component of the war effort: Bullets manufactured in one state, for example, had to work consistently in rifles made in another. The armed forces initially inspected virtually every unit of product; then to simplify and speed up this process without compromising safety, the military began to use sampling techniques for inspection, aided by the publication of military-specification standards and training courses in Walter Shewhart's statistical process control techniques.

The birth of total quality in the United States came as a direct response to the quality revolution in Japan following World War II. The Japanese welcomed the input of Americans Joseph M. Juran and W. Edwards Deming and rather than concentrating on inspection, focused on improving all organizational processes through the people who used them.

By the 1970s, U.S. industrial sectors such as automobiles and electronics had been broadsided by Japan's high-quality competition. The U.S. response, emphasizing not only statistics but approaches that embraced the entire organization, became known as total quality management (TQM).

By the last decade of the 20th century, TQM was considered a fad by many business leaders. But while the use of the term TQM has faded somewhat, particularly in the United States, its practices continue.

In the few years since the turn of the century, the quality movement seems to have matured beyond Total Quality. New quality systems have evolved from the foundations of Deming, Juran and the early Japanese practitioners of quality, and quality has moved beyond manufacturing into service, healthcare, education and government sectors.

The terms "quality assurance" and "quality control" are often used interchangeably to refer to ways of ensuring the quality of a service or product. The terms, however, have different meanings.

Assurance: The act of giving confidence, the state of being certain or the act of making certain.

Quality assurance: The planned and systematic activities implemented in a quality system so that quality requirements for a product or service will be fulfilled.

Control: An evaluation to indicate needed corrective responses; the act of guiding a process in which variability is attributable to a constant system of chance causes.

Quality control: The observation techniques and activities used to fulfill requirements for quality.

In simple yet profound terms, variation represents the difference between an ideal and an actual situation. An ideal represents a standard of perfection–the highest standard of excellence–that is uniquely defined by stakeholders, including direct customers, internal customers, suppliers, society and shareholders. Excellence is synonymous with quality, and excellent quality results from doing the right things, in the right way.

The fact that we can strive for an ideal but never achieve it means that stakeholders always experience some variation from the perfect situations they envision. This, however, also makes improvement and progress possible. Reducing the variation stakeholders experience is the key to quality and continuous improvement.

According to the law of variation as defined in the *Statistical Quality Control Handbook:*

- "Everything varies." In other words, no two things are exactly alike.
- "Groups of things from a constant system of causes tend to be predictable." We can't predict the behavior or characteristics of any one thing. Predictions only become possible for groups of things where patterns can be observed.

If outcomes from systems can be predicted, then it follows that they can be anticipated and managed.

Managing Variation

In 1924, Dr. Walter Shewhart of Bell Telephone Laboratories developed the new paradigm for managing variation. As part of this paradigm, he identified two causes of variation:

- Common cause, or noise, variation is inherent in a process over time. It affects every outcome of the process and everyone working in the process. Managing common cause variation thus requires improvements to the process.
- Special cause, or signal, variation arises because of unusual circumstances and is not an inherent part of a process. Managing this kind of variation involves locating and removing the unusual or special cause.

Shewhart further distinguished two types of mistakes that are possible in managing variation: treating a common cause as special and treating a special cause as common. Later, W. Edwards Deming estimated that a lack of an understanding of variation resulted in situations where 95 per cent of management actions result in no improvement. Referred to as tampering, action taken to compensate for variation within the control limits of a stable system increases, rather than decreases, variation.

"The Cost of Quality"

It's a term that's widely used – and widely misunderstood. The "cost of quality" isn't the price of creating a quality product or service. It's the cost of NOT creating a quality product or service.

Every time work is redone, the cost of quality increases. Obvious examples include:

- The reworking of a manufactured item.
- The retesting of an assembly.
- The rebuilding of a tool.
- The correction of a bank statement.
- The reworking of a service, such as the reprocessing of a loan operation or the replacement of a food order in a restaurant.

In short, any cost that would not have been expended if quality were perfect contributes to the cost of quality.

Total Quality Costs

As the figure below shows, quality costs are the total of the cost incurred by Investing in the prevention of non conformance to requirements, Apparising a product or service for conformance to requirements, Failing to meet requirements

Quality Costs – General Description

Prevention Costs

The costs of all activities specifically designed to prevent poor quality in products or services.

Examples are the costs of:

- New product review
- Quality planning

- Supplier capability surveys
- Process capability evaluations
- Quality improvement team meetings
- Quality improvement projects
- Quality education and training

Appraisal Costs

The costs associated with measuring, evaluating or auditing products or services to assure conformance to quality standards and performance requirements.

These include the costs of:

- Incoming and source inspection / test of purchased material
- In-process and final inspection / test
- Product, process or service audits
- Calibration of measuring and test equipment
- Associated supplies and materials

Failure Costs

The costs resulting from products or services not conforming to requirements or customer / user needs. Failure costs are divided into internal and external failure categories.

Internal Failure Costs

Failure costs occurring prior to delivery or shipment of the product, or the furnishing of a service, to the customer.

Examples are the costs of:

- Scrap
- Rework
- Re-inspection
- Re-testing
- Material review
- Downgrading

External Failure Costs

Failure costs occurring after delivery or shipment of the product — and during or after furnishing of a service — to the customer.

Examples are the costs of:

- Processing customer complaints
- Customer returns
- Warranty claims
- Product recalls

Total Quality Costs

The sum of the above costs. This represents the difference between the actual cost of a product or service and what the reduced cost would be if there were no possibility of substandard service, failure of products or defects in their manufacture.

11

Packaging Testing

Paper Based

Paper based materials are bought on the basis of weight. Hence the weight of the packaging material forms a primary parameter to assess the quality. In addition there are other parameters which reflect the quality and there by the performance of the packaging material. Given below is a list of tests that needs to be conducted on paper based material and it has been categorized in the terms of forms of packaging for easy reference. All these tests are conducted as per standards. It is needless to mention that packaging materials before testing need to be conditioned to Indian standard conditions of 27°C and 65 per cent RH for atleast 24 hours. This is required because testing is basically conducted to find the suitability of material for particular function and hence testing of conditioned sample will ensure values under standard base conditions.

Paper and Paper Board

1. Grammage
2. Thickness
3. Cobb
4. Bursting Strength
5. Burst Factor
6. Gloss
7. Machine Direction and Cross Direction
8. Tensile Strength and Elongation

9. Opacity
10. Brightness
11. Smoothness and Porosity
12. Moisture
13. Ash content
14. Chloride Content
15. Sulphide content

Corrugated Fibre Board

1. Grammage
2. Individual grammage
3. Bursting Strength
4. Burst Factor of paper
5. Puncture Resistance
6. Moisture
7. Flute profile
8. Edge Crush test
9. Flute height
10. Take up factor
11. Dimensions
12. Style of the box
13. Grade of paper
14. Adhesive content
15. Drop test
16. Rolling test
17. Vibration
18. Inclined Impact
19. Compression Strength

Paper Bags

1. Grammage
2. Thickness
3. Cobb
4. Bursting Strength
5. Burst Factor
6. Machine Direction and Cross Direction
7. Tensile Strength and Elongation

8. Moisture
9. Dimensions
10. COF
11. Drop test
12. Rolling test
13. Vibration
14. Inclined Impact
15. Compression Strength

Paper Board Cartons

1. Grammage
2. Thickness
3. Cobb
4. Bursting Strength
5. Burst Factor
6. Machine Direction and Cross Direction
7. Stiffness
8. Opacity and Brightness
9. Scuff Proofness
10. Dimensions
11. Drop test
12. Rolling test
13. Vibration
14. Inclined Impact
15. Compression Strength

Polymer Based

Polymer based materials are bought on the basis of weight, and used in numbers or area. Hence yield of plastic material will form important parameter along with weight of the packaging material. In addition there are other parameters which reflect the quality and there by the performance of the packaging material. Given below is a list of tests that needs to be conducted on paper based material and it has been categorized in the terms of forms of packaging for easy reference. All these tests are conducted as per standards. It is needless to mention that packaging materials before testing need to be conditioned to Indian standard conditions of 27°C and 65 per cent RH for at least 24 hours. This is required because testing is basically conducted to find the suitability of material for particular function and hence testing of conditioned sample will ensure values under standard base conditions.

Bottles and Cap

1. Bottle dimensions
2. Wall thickness
3. Thread Profile
4. Verticality
5. Migration
6. ESCR
7. Weight
8. Drop test
9. Rolling test
10. Vibration
11. Inclined Impact
12. Compression Strength

Lamines, Laminated Bags, Films and Film Bags

1. Material Identification
2. Tensile strength
3. Tear Strength
4. Grammage
5. Individual Grammage
6. Seal Strength
7. Bond Strength
8. COF
9. WVTR
10. OTR
11. Migration
12. Thickness
13. Gloss
14. Opacity and Brightness
15. Scuff Proofness
16. Drop test
17. Rolling test
18. Vibration
19. Inclined Impact
20. Compression Strength

HDPE Woven Sacks

1. Weight
2. Grammage
3. Denier
4. Mesh
5. Material Identification
6. Seal Strength
7. Design
8. Dimension
9. Tear Strength
10. Tensile strength and Elongation
11. WVTR
12. OTR
13. Stitching type
14. Drop test
15. Rolling test
16. Vibration
17. Inclined Impact
18. Compression Strength

HDPE Drugs

1. Dimensions
2. Wall thickness
3. Mouth orifice dimensions
4. Drop test
5. Rolling test
6. Vibration
7. Inclined Impact
8. Compression Strength

Wood, Jute and Cotton Based

Wood

1. Moisture content
2. Nail holding power
3. Design
4. Dimensions
5. Drop test

6. Rolling test
7. Vibration
8. Inclined Impact
9. Compression Strength

Jute Bags and Cotton Bags

1. Dimension
2. Design
3. Tensile Strenth and Elongation
4. GSM
5. Denier
6. Mesh
7. Stitching Style
8. No. of Stitches
9. Sealing strength
10. Trapezoidical Tear
11. Drop test
12. Rolling test
13. Vibration
14. Inclined Impact
15. Compression Strength

Metal Based

MS Drums

1. Dimensions
2. Wall thickness
3. Mouth orifice dimensions
4. Drop test
5. Rolling test
6. Vibration
7. Inclined Impact
8. Compression Strength

Tin Plate Container

1. Migration
2. Thickness of plate
3. Hardness

4. Profile of container
5. Beading dimensions
6. Dimensions
7. Lacquer coating
8. Drop test
9. Rolling test
10. Vibration
11. Inclined Impact
12. Compression Strength

Simulated Package Testing

Testing of Packages

Functional Requirements

The basic function of a package is to protect and preserve the contents during transit from the manufacturer to the consumer. Protection is required against spillage, dirt, ingress and egress of moisture, insect infestation, contamination by foreign material, tampering, pilferage, etc. A package should preserve the contents in ' factory-fresh' condition during the period of storage and voyage by ensuring freedom from bacteriological attack, chemical reaction, etc. An important distinction is to be made here between two types of packaging: Transit Packaging and Consumer Packaging.

Transit Packaging

Most of the products entering the export trade require an outer package of one kind or another. An ideal transport package is one which offers protection against loss and damage during handling, transportation and storage and is at the same time, light and economical. The choice of the package, whether a wooden crate, a fibre board box or any other type – is to be governed by the characteristics of the product on the one hand and on the other, the kind of handling and transportation hazards likely to be encountered at various stages, not only in the exporting country but also during transit in the importing country.

A transit package is mainly expected to offer protection against handling and transportation hazards while a consumer that the product reaches the ultimate consumer in prime condition, free from contamination by dirt, moisture, heat, insects, etc. during transit and storage.

The best package for any particular purpose is the one which would protect the contents against the hazards the package would undergo during its journey at the minimum cost. The simplest and the most efficient way of testing packages is to carry out field trials with sufficient number of packages under the actual conditions of usage. However, such field trials are time consuming and require a fairly large number of packages sent out on several occasions by different routes.

Evaluation of package performance or package testing is a means of shortening this process and of obtaining results in a shorter period with a reasonable degree of accuracy.

There are four main hazards of transport:

i) Drops and Impacts
ii) Compression forces
iii) Vibration and Vibration under stacking loads
iv) Climatic variations

Equipments are available for testing packages for all these hazards.

Mechanical Test

Drop Test

This test helps to measure the ability of the container and inside packing materials to provide protection to its contents, and to measure the ability of the container information useful improving the design of the container.

The common types of apparatus are available, *viz.* the Divided table top Drop test apparatus and the Hoist type with suitable slings, tripping device and hooks. In the first type of Tester, (which is very useful with packages which are difficult to sling) called the Table drop Tester, the package is held or supported in the desired position of fall and the trap door opened, thereby projecting the package on to the floor. The height of drop. The position of fall and the type of floor can be altered at will.

The second type of tester consists of some form of release mechanism from which the package is suspended by means of a sling which allows it to be dropped in any particular position onto any type of floor from any selected height.

The drop test may be divided into two procedures. Procedure – A shall be used for measuring the ability of the container to provide protection to its contents. In this procedure the drop test included cornerwise, edgewise and flat-wise drops. Procedure-B shall be used for measuring the ability of the container to with stand rough handling. In this procedure the drop test generally consists of cornerwise drops.

The drop test results help to conclude the adequacy of packing and to ascertain the relative ability of the container to with stand the normal hazards encountered by a container from the time it is packed to the time it is opened for use.

The information which may be obtained from this test is of great value to the shipper and the container manufacturer. It is frequently possible, through the analysis of results to reduce the cost of package, or to improve the design of the package to prelude damage in transit. It affords the manufacturer of product a knowledge that the design of the product either does or does not, lend itself to economical packaging.

Vibration Test

The test is conducted to determine:

(i) The ability of the container

(ii) The protection offered by materials used for interior packing, and

(iii) The strength of the closures used

Vibration shocks could be encountered both during inplant handling and in shipment. Internal handling on conveyers and industrial trucks could produce damage to such highly fragile products as electronic components. However, the vast majority of vibration damage occurs in transit.

The effects of vibration are:

a) Structural failure in packaging materials through vibration
b) Reduction in container rigidity through weaving and swaying usually aggravated by wracking and skewing if the container bears a top load.
c) Exterior and interior damage through abrasion and
d) Damage to contents

A vibrating table is used for studying this type of hazard. The vibrating table consists of a bed which is driven by two eccentrics, one at each end, connected in phase with one another. To the top of the vibrating bed a platform is attached, and the platform describes a circular harmonic type of vibration when the equipment is running. The amplitude of the vibration on this instrument is fixed to one inch, and the frequency may be varied continuously from about 120 cycles a minute to about 360 cycles a minute.

This test helps to study the possible package failure of content damage, and the weakness (loosening of bonds, separation of container layer, abrasion of smooth surfaces, loss of resilience in cushioning material etc.) that may develop during transit.

Compression Test

This test is carried out, generally, on empty containers to measure the ability of the container to resist external compressive loads applied to faces and when applied to diagonally opposite edges or corners.

Compression stresses may be caused by (1) static condition due to the superimposition of weight onto merchandise of by (2) dynamic stresses of impact stresses. While the static impression is primarily experienced in stacking, the dynamic compressive stresses generally result from impact shock in handling and transit.

During storage or transport, packages are stacked to the maximum height possible. Consequently the bottom layer is subjected to compression and hence has to bear the load of the packages stacked over it. If the container is not strong enough the force of compression is transmitted to the contents. This is particularly so, in case of fibre board containers and the currently used bamboo baskets for fruits and vegetables.

Thus, the container used should have adequate compression strength for a given purpose. This is determined by means of standard compression tester. The data obtained would indicate the strength of the container for short period of stacking, for long period it is preferable to take the compression strength as one third of the test value. For dynamic conditions the performance strength could be taken as the test value multiplied by 3/2 in the vertical direction.

Inclined Impact Test (Contour Test)

This test helps to study the extent of crushing, breaking, cracking, distortion and shifting during handling storage and transport. These are mainly due to shunting shocks. This test is carried out to determine the ability of a container to withstand impact stresses and also to determine the ability of a container or interior packing, or both, to provide protection to the contents when subjected to impact stresses.

This test is particularly valuable for testing large, odd shaped or heavily loaded containers that would be difficult or impossible to test by other methods.

The study is carried out generally with an Inclined-plane Impact Tester. This consists of a track inclined at 10 per cent to the horizontal, on which a dolly can be released so as to impact at the wooden buffers placed at right angles across the end. According to the distance of the incline from which the dolly is released, the impact speed can be varied upto about 8 m.p.h. The results of the tests, are indicative of the effects experienced by the package when it is subjected to shunting shocks of side impacts during transit.

From the results obtained, it is possible to design suitable containers or to improve the design of the existing containers or interior packing.

Rolling Test

The rolling test is performed as follows:

The package, standing on its base, is slowly pushed to rotate about one edge of the base and allowed to fall over on the side face. After this impact, the package is lifted, balanced on its edge and then allowed to fall over on the top. The procedure is repeated until the package once again stands on its base.

The whole sequence from the base to base traveling in the direction at a right angle to that taken in the first half of the operation is repeated.

This test helps to evaluate the overall strength of the container and the cushioning provided inside, and any failure of the contents.

Results obtained from this test help to design a suitable container, assess the adequacy of the cushioning material and to arrive at suitable and economical placement of the contents in the bulk container.

Climatic Tests

Rain Test

A simulated rainfall of 4x1 inches per hour should be provided by a water spray nozzle designed to emit water in small droplets rather than a fine mist. The rainfall should be dispersed uniformly over the test area. Duration of the test is generally two hours.

Sand and Dust Tests

The purpose of sand and dust tests is to evaluate the resistance of a package to the penetration of sand and dust, to determine the erosive effects of blowing sand and dust on exposed items and to determine the immediate effects such as the malfunction of moving parts, by the presence of sand and dust. A standardized mixture of sand and dust of density 0.1 to 0.5 gm / cu. Ft. is used to create an atmosphere for this. The temperature of this atmosphere is maintained at 77°F for a period of 6 hours and then increased to 160°F for another 16 hours. A standard sand and dust velocity of 200′ per min. is maintained throughout the test unless an optional velocity of 2,300 ft. per min. is specified. Items which are always shielded from blowing sand and dust should be tested under standard velocity conditions, 200′ per min. Items which may be exposed to blowing sand and dust should be tested under optional velocity conditions 2,300 per min.

Salt Spray Test

The purpose of the salt spray test is to evaluate the resistance of the package to corrosion by salt spray and to serve as a general standard for corrosion resistance. No correction exists between the test time and the natural exposure time.

The package is exposed for 50 hours, to a wet, dense fog generated by the automation of a 20 per cent water solution of sodium chloride. The solution shall be maintained at a pH of between 6.5 to 7.2, the temperature of the fog being maintained at 95°F.

Fungus Resistance Tests

All materials used in the fabrication of shipping containers should be tested for fungus resistance.

A composite spore suspension is prepared as follows:

10 ml of distilled water containing approximately 0.005 per cent dioctyl sodium sulfacuccinate is introduced directly into each culture. The culture is then raked with a sterile wire and agitated to insure a well sporated suspension. The test item is then placed in a humidity of 25 per cent and a temperature of 86°F for 28 days. If, at the end of 28 days, 25 per cent or less of the samples or test items have fungus growth on more than 4 per cent of their exposed surface, the material is considered to be fungus nutrient. If 2 per cent or less of the exposed surface is covered, the material is considered to be inert for fungus growth.

The foregoing gives only a brief idea of the possible laboratory and field tests. These tests give the following correlations to obtain a satisfactory picture on package performance. A correlation between the performance of the package in service and the laboratory transport test which reproduces in some degree the events which would happen to the package in the field. Such a laboratory transport test is capable of evaluating a package as to its probable efficiency. The user of the package is primarily concerned with this correlation.

A correlation between the laboratory transport test on the package and tests on the empty container and any other component such as cushions etc. thus would enable the package maker to use simpler control tests on bulk production.

The third correlation to be made between the materials of construction and the containers.

- The particular type of testing used in any instance will be dependent on the circumstances and the information desired.
- In general, it might be said, that all these tests would help to conclude:
  - The overall durability of the container
  - Any possible reduction in the cost of packaging
  - Improvement required in the design of the box to preclude damage in transit
  - Adequacy of the cushioning used
- In short, a series of tests would enable to determine the transport worthiness of the package.

12

Modern Food Packaging Techniques

Active Packaging

Definition

Active when the packaging elements change the condition of the packed food to extend shelf-life or improve safety or sensory properties, while maintaining quality of packaged food. To understand what active and intelligent packaging have to offer the world of packaging, it is important to clarify what each phrase means. Active packaging is accurately defined as "packaging in which subsidiary constituents have been deliberately included in or on either the packaging material or the package headspace to enhance the performance of the package system". This phrase emphasizes the importance of deliberately including a substance with the intention of enhancing the food product. Active packaging is an extension of the protection function of a package and is commonly used to protect against oxygen and moisture

Active packaging has been investigated for more than 40 years, or ever since passive packaging embracing oxygen and water vapour barriers became important to the protection of food and beverage products during distribution, that trek from the end of the production line to the consumer.

The main purpose of food packaging is to protect the food from microbial and chemical contamination, oxygen, water vapour and light. The type of packaging used therefore has an important role in determining the shelf life of a food. 'Active' packaging does more than simply provide a barrier to outside influences. It can control, and even react to, events taking place inside the package. Fresh foods just after harvest or slaughter are still active biological systems. The atmosphere inside a package constantly changes as gases and moisture are produced during metabolic processes. The type of packaging used will also influence the atmosphere around the food because some plastics have poor barrier properties to gases and moisture. The metabolism of fresh food continues to use up oxygen in the headspace of a package and increases the carbon dioxide concentration. At the same time water is produced and the humidity in the headspace of the package builds up. This encourages the growth of spoilage micro organisms and damages the fruit and vegetable tissue. Many food plants produce ethylene as part of their normal metabolic cycle. This simple organic compound triggers ripening and aging. This explains why fruit such as bananas and avocados ripen quickly when kept in the presence of ripe or damaged fruits in a container and broccoli turn yellow even when kept in the refrigerator. Extensive trials have shown that each fresh food has its own optimal gas composition and humidity level for maximizing its shelf life. Active packaging offers promise in this area; it is difficult with conventional packaging to optimise the composition of the headspace in a package. The atmosphere surrounding the food also influences the shelf life of processed foods. For some processed foods, a lowering of oxygen is beneficial, slowing down discoloration of cured meats and powdered milk and preventing rancidity in nuts and other high fat foods. High carbon dioxide and low oxygen levels can pose a problem in fresh produce leading to anaerobic metabolism and rapid rotting of the food. However, in fresh and processed meats, cheeses and baked goods, carbon dioxide may have a beneficial antimicrobial effect. Active packaging employs a packaging material that interacts with the internal gas environment to extend the shelf life of a food. Such new technologies continuously modify the gas environment (and may interact with the surface of the food) by removing gases from or adding gases to the headspace inside a package. Recent technological innovations for control of specific gases within a package involve theuse of chemical scavengers to absorb a gas or alternatively other chemicals that may release a specific gas as required. The table below sets out some areas of atmosphere control in which active packaging is being successfully used.

Active Packaging Applications in the Food Industry

Types of Active Packaging Application	
Absorbing/scavenging properties	Oxygen, carbon dioxide, moisture, ethylene, flavors,taints, UV light
Releasing/emitting properties	Ethanol, Carbon dioxide, antioxidants, preservatives,sulfur dioxide, flavors, pesticides
Removing properties	Catalyzing food component removal (lactose, cholesterol)
Temperature control	Insulating materials, self-heating and self-coolingpackaging, microwave susceptors and modifiers,temperature-sensitive packaging
Microbial and quality control	UV and surface treated packaging materials

Active Packaging	System Application
Oxygen scavenging	Most food classes
Carbon dioxide production	Most food affected by moulds
Water vapour removal	Dried and meld-sensitive foods
Ethylene removal	Horticultural produce
Ethanol release	Baked foods (where permitted)

Intelligent Packaging

Intelligent packaging can be defined as "packaging that contains an external or internal indicator to provide information about aspects of the history of package and/or the quality of the food". Intelligent packaging is an extension of the communication function of Traditional packaging, and communicates information to the consumer based on its ability to sense, detect, or record external or internal changes in the product's environment. Intelligent packaging systems exist "to monitor certain aspects of a food product and report information to the consumer." The purpose of the intelligent system could be to improve the quality or value of a product, to Provide more convenience, or to provide tamper or theft resistance. Intelligent packaging can report the conditions on the outside of package, or directly measure the quality of the food product inside the package. In order to measure product quality within the package, there must be direct contact between the food product or headspace and the quality marker. In the end, an intelligent system should help the consumer in the decision making process, to extend shelf life, enhance safety, improve quality, provide information and warn possible problems. Intelligent packaging is a great tool for monitoring possible abuse that has taken place during the food supply chain. Intelligent packaging may also be able to tell a consumer when a package has been tampered with. There is currently work being developed with labels or seals that are transparent until a package is opened. Once the package is tampered with, the label or seal will undergo a permanent color change and may even spell out "opened" or "stop". Perhaps intelligent packaging will be able to inform a consumer of an event that occurred such as package tampering that may save their life

Time Temperature Indicators (TTIs)

The TTI is useful because it can tell the consumer when foods have been temperature abused. If a food is exposed to high temperature than recommended the quality of food can deteriorate much quicker.

Intelligent Packaging: Gas Indicators

Gas indicators are a helpful means of monitoring the composition of gases inside a package by producing change in colour of the indicator through enzyme or chemical reaction.

Intelligent Packaging: Thermochromics Inks

Inks are available that are temperature sensitive and can change colors based on temperature.

These inks can be printed onto packages or labels such that the message can be conveyed to consumer based on colour of ink they can see

Modified Atmosphere Packaging (MAP)

Modified atmosphere packaging can be used to extend the shelf life of many fruit and vegetables. This technology seems straightforward as it uses permeable films and the respiration rate of the product at a specific temperature to change the concentration of carbon dioxide and oxygen around the product. The main aim of modified atmosphere packaging (MAP) is to change the composition of the atmosphere around the product so that the storage life of the product can be extended. Most fruit and vegetables age less quickly when the level of oxygen in the atmosphere surrounding them is reduced. This is because the reduced oxygen slows down the respiration and metabolic rate of the products and therefore slows down the natural aging process. Raising the level of carbon dioxide to levels of 2 per cent or more can also be beneficial. Elevated CO_2 levels can reduce the products sensitivity to ethylene, it can also slow the loss of chlorophyll which is the green colour of fruit and vegetables. High CO_2 can also slow the growth of many of the postharvest fungi that cause rots. All these effects can help to extend the storage and shelf life of fresh produce. However, while basically a simple system for commercial usage it is critical that considerable care is paid to several factors the most important of which is temperature control.

The Theory of MAP

When a given weight of produce is sealed within a plastic bag, it uses oxygen and produces carbon dioxide. As the oxygen concentration inside the package falls, below about 10 per cent the rate of respiration (oxygen use) starts to decrease. At the same time, oxygen moves into the bag through the walls of the plastic bag and carbon dioxide moves out. Oxygen and carbon dioxide move across the film in proportion to the drop in concentration of oxygen and rise of carbon dioxide concentration inside the plastic bag.

This seems simple however the rate of oxygen consumed is dependent on the following factors;

- The weight of the product in the bag
- The temperature and
- The respiration rate of the commodity. Respiration rate may vary among cultivars, seasons and growing conditions.
- The rate of oxygen and carbon dioxide movement through the wall of the bag

The rate of oxygen movement through the plastic bag depends on the **surface area, thickness and chemical properties of the plastic film**. The permeability of the film can be increased by adding holes. Commercially microporous films, that has a number of tiny holes ensures enough oxygen is supplied to the product when this film is used. The difficulty with using Modified Atmosphere Packaging is the establishment of a stable atmosphere inside the plastic bag. MAP is a dynamic system that is not controlled. As currently used there is no feedback system that can cut inif one of the factors listed above changes. Therefore it is important to use MAP packaging only as recommended by the manufacturer of the packaging material.

The factor that causes most problems in a commercial situation is temperature. Unfortunately the cool chain for fresh produce is not always continuous throughout the marketing system. Breaks in the cool chain such as during loading or unloading of trucks or packing of warehouses mean that the cool product can warms up. Warming of only a few degrees can be enough to cause the respiration rate of the product to rise and the oxygen within the package to fall below the recommended level. If the oxygen level falls too low then anaerobic respiration can be initiated. If this happens alcoholic off flavours develop within the product, making it unmarketable. There is always a risk/benefit when using modified atmosphere packaging, particularly when a low oxygen atmosphere is providing the benefit. The greatest extension of shelf life occurs at the lowest possible oxygen concentration before anaerobic respiration is initiated. This point also carries the greatest risk. For example, if the respiration rate increases as a result of a small change in temperature then the oxygen level will fall below the critical level and off flavours will be produced. The same is true for atmospheres where the main benefit is high carbon dioxide. If

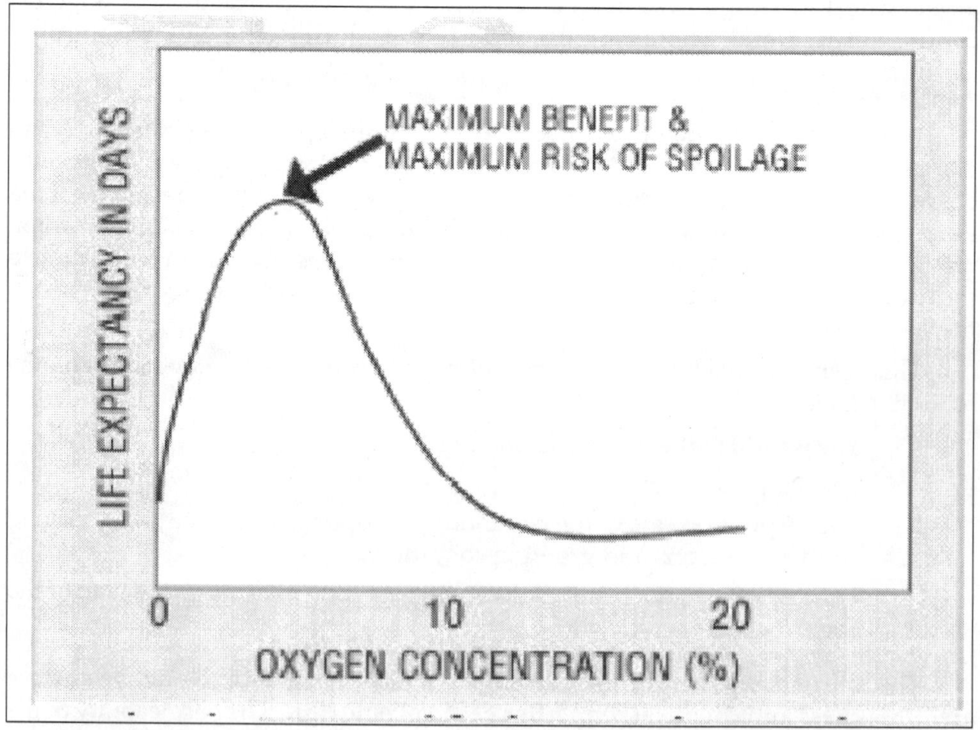

For the Risk/Benefit Associated with using Modified Atmosphere Packaging.

respiration increases due to an increase in temperature then the level of CO_2 may rise above the critical level and the product will also be damaged and made unsaleable. There are two ways to minimise the risk of spoilage. Firstly, use a package that provides slightly more oxygen, and so provides less benefit in terms of shelf life but the package would also have a reduced risk of spoilage. Secondly, ensure that the cool chain is maintained. If temperature cannot be guaranteed then it is a very great risk with this type of packaging system. If the temperature rises by more than a few degrees then damage could be avoided by opening the bags to ensure adequate oxygen for the product. This is not often feasible but some packers recommend that the MA bags are opened once the product arrives at the wholesale market to ensure there is no risk of spoilage. The following diagram illustrates this point. Modified atmosphere packaging is a cheap and convenient packaging system that has the capacity to extend the shelf life of some commodities if it is used properly. If the complexity of the MA packaging is understood then it is more likely that the product will be handled with the proper care. The risks of this seemingly simple packaging system must always be considered.

Controlled Atmosphere Packaging (CAP)

The use of controlled atmospheres for extending the life of fresh produce, stored in bulk, is well known concept developed around 60 years ago. Controlled atmosphere storage, in rooms where temperature and atmosphere are carefully controlled, is largely responsible for the year round availability of fresh fruits such as apples and pears. More recently, the concept has been extended to an ever widening circle of fresh products - tomatoes, cabbage, lettuce, strawberries and fresh meats. More so, it has moved fresh products at both the wholesale and retail level from bulk storage to controlled atmosphere packaging. The terms "controlled atmosphere" and "modified atmosphere" have been used in different ways by different authors. To these, one can add technologies such as vacuum packaging, shrink wrapping or even wax or other coatings that will change, modify or otherwise "control" the macro- or micro atmosphere surrounding the fresh product. Here, the term "controlled atmosphere" will be used with the context clarifying the specific technology being employed. The "atmosphere" being controlled is that in direct contact with the surface of the fresh product, and includes attempts to control oxygen, nitrogen, carbon dioxide, water vapour, and temperature. The reason for wishing to control these particular variables is that they affect respiration of the product directly. Properly controlled, they can slow down normal aging processes. For example, raising the carbon dioxide level and lowering the oxygen level **can slow**, but not necessarily, nor desirably, stop, normal respiration. The relative humidity around the product does not have a major effect on respiration, but is crucial in controlling transpiration. Carbon dioxide slows, or eliminates the growth of certain moulds and other microorganisms. The objective of this atmospheric control is to extend product shelf life. The emphasis is on Controlled Atmosphere Packaging, where the end product is a packaged atmosphere that is in equilibrium with the external atmosphere.

Features of Controlled Atmosphere Packaging (CAP)

Significant points in the situation of controlled atmosphere packaging can be summarized as follows:

(a) The objective of controlled atmosphere packaging for fresh products is to **prolong high quality, not to improve or enhance quality**. Thus the onset of, or rate of, oxidation, putrefaction, etc. is slowed, but the final product will exhibit a diminished quality when compared to the freshly picked or prepared product, even though the quality loss may be negligible. There may be some beneficial effects in taste or texture due to dissolution of carbon dioxide into moisture at the tissue surface. Also, the absorption of carbon dioxide will lower pH, reducing the opportunity of harmful organisms.

(b) Since the usual objective of controlled atmosphere packaging is to slow the natural respiration rate of the fresh product, though not to stop it altogether, controlled atmospheres will generally cause lowered oxygen levels, elevated carbon dioxide levels and lowered temperatures. Controlled atmospheres sometimes work well in combination with low temperatures, while in other cases, they may replace low temperature storage as a means of extending shelf life.

(c) Successful controlled atmosphere packaging should result in either an equilibrium between the respiration rate of the product and the influx and escape of gases through the package, or, at least, a predictable rate of change of the levels of different gases inside the package.

(d) Since the results required in (c) above are strongly dependent upon temperature, closely controlled storage temperatures the upper temperature limits are mandatory for effective controlled atmosphere packaging. Whereas in conventional processed food packaging, potentially harmful micro-organisms and parasites are either killed through heat processing or rendered inert by freezing, in controlled atmosphere packaging, the only micro-organism control possible is through sanitation and use of fungicides or bactericides during harvest, preparation and packaging. This can be a somewhat haphazard and unreliable approach, as storage conditions may actually provide a breeding ground for micro-organisms.

(e) Generally, only the macro-environment within the package can be controlled, and there is a possibility that different micro-environments can exist. For example, an anaerobic condition can occur deep within the product itself, even if the controlled environment in the package is aerobic and hostile to certain pathogens.

(f) Final condition of the package and its controlled atmosphere can only be determined safely by careful studies of the packaged product, including microbiological determinations and sensory testing panel evaluations.

(g) In order to maximize the potential for extending retail shelf life of fresh food products, it is likely that major changes to the distribution chains for these products will be required.

Vacuum Packaging

Vacuum packaging refers to packaging in containers (rigid or flexible), from which substantially all air has been removed prior to final sealing of the container. This method of packaging is actually a form of "Modified Atmosphere" since normal room air is removed from the package.

Advantages of Vacuum Packaging

1. Extends shelf life - Aids in controlling oxidative rancidity.Prevents the growth of normal spoilage bacteria.Aerobic organisms such as Psuedomonas are suspended and lactic-acid bacteria are favored. The latter can grow to high numbers without causing spoilage.
2. Reduces moisture loss and freezer burn - Prevents movement of water out of the product into the surrounding headspace.Prevents loss of moisture at product surface and eliminates freezer burn.
3. Requires minimal storage space - Package is drawn tight around product taking up minimal space.
4. Leakers are easily detected - A small puncture or pinhole in a vacuum pack is easy to detect by looking for loose packages.

Extruded Foods/Canning and Retort Packs

Extrusion is a process, which combines several unit operations including mixing, cooking, kneading, shearing, shaping and forming. In essence an extruder consists of a screw pump. This forces the dough through a restricted opening (the Die) at the discharge end of the screw. If the food is heated the process is known as extrusion cooking (or Hot extrusion). The main purpose of extrusion is to increase the variety of foods in the Diet, by producing a range of products with different shapes, texture, colours and flavours from basic ingredients. Extrusion cooking is a high temperature short time (HTST) process, which reduces microbial contamination and inactivates enzymes. However, the main method of preservation of both hot and cold-extruded food is by the low water activity of the product. Extrusion is gaining in popularity for the following reasons.

- Versatility: Wide range of products by managing slight changes.
- Reduced cost.
- High production rate and Automatically production.
- No process effluents are produced.

Theory: During extrusion cooking of the starch based foods (*i.e.* Maize Grit and wheat flours); the moisture content is increased by added water and the starch is subjected to intense shearing forces at an elevated temperature. The starch granules swell and absorb water and becomes gelatinized and a viscous plasticized mass is produced. This is extruded *e.g.* fruit gum liquorice. During Protein based goods (*i.e.* soy meal and defatted oil-seed flours) the quaternary structure of protein opens in the hot moist conditions; to produce a viscous plasticized mass. The proteins are then polymerized; cross-linked and reoriented to form the fibrous structure of texturised vegetable protein (TVP).

This is the extruded through machine. *e.g.* cornflakes.

Products

- Noodles
- Corn flakes
- Fruit Gum
- Liquorice
- Precooked rice flakes
- Cooked potato cubes

Asepctic – Packing

"Packing of sterile fruit pulp" into separately sterile packaging under sterile conditions". The detailed production pathway is explained below.

The packing materials are "foil based" for longer shelf life and "Metalized – polyester" based for comparatively shorter shelf life, for mainly fruit pulp/puree. The "Tetra-pack" used for juice packing is "foil-paper" based.

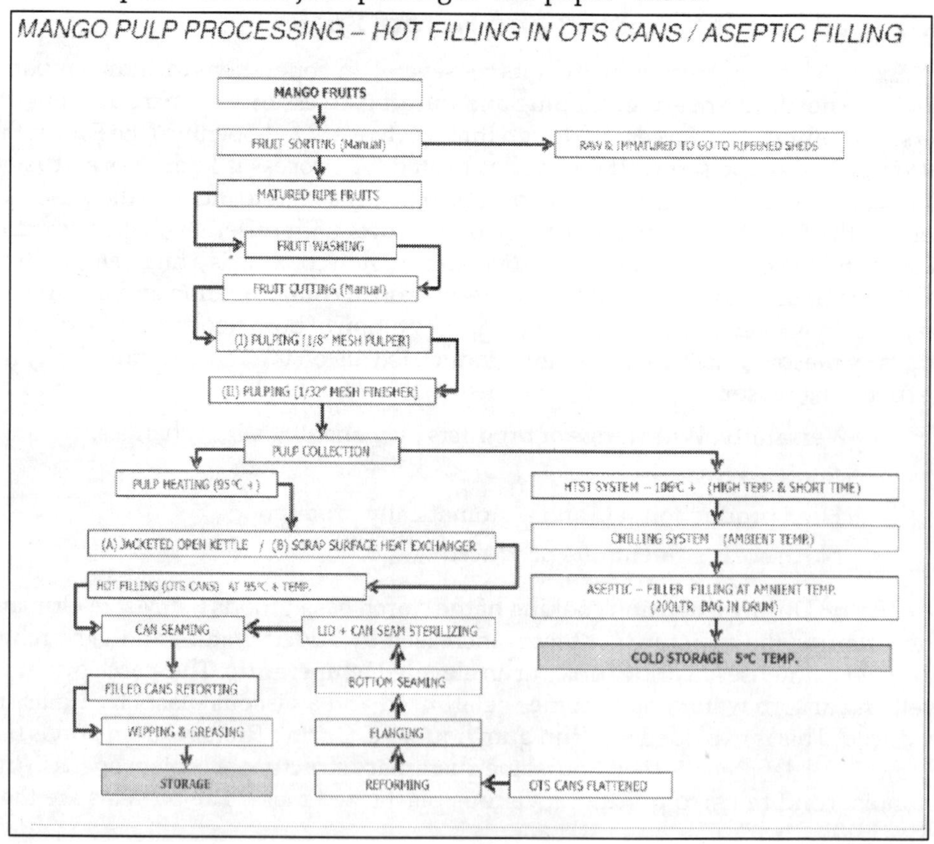

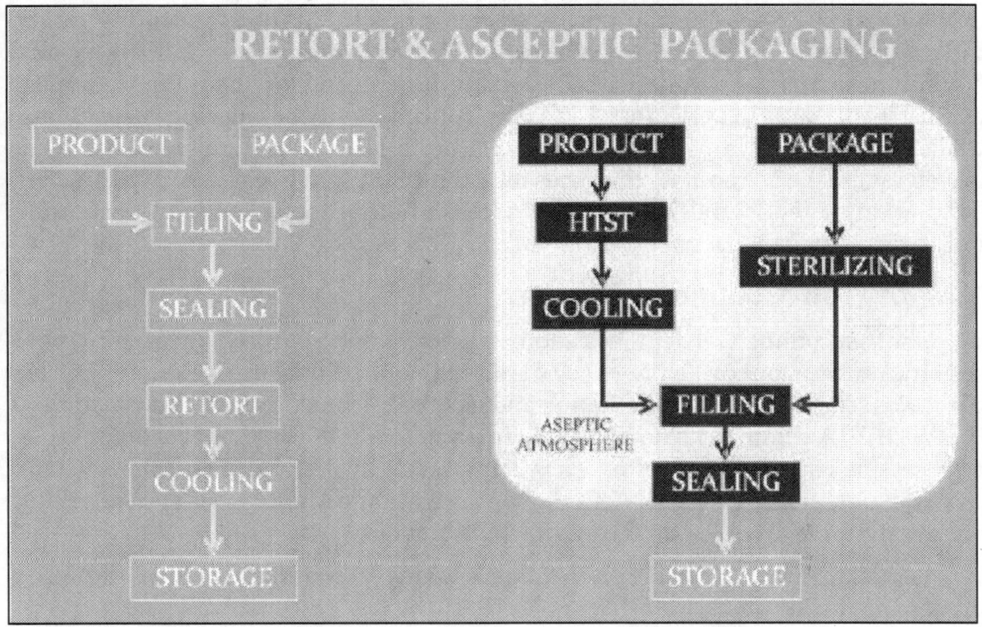

Retort Pouch/Flexi-Cans

Packaging Compositions

1. Polyester / Alu-foil / cast polypropylene
2. Polyester / Alu-foil / modified high-density polyethylene.

Processing

Similar to OTS cans hot fillings canning process, retort pouch is filled-hot and sealed. It passes through retort chamber with counter pressure maintained in chambers. When pouches with product packed goes to hot retort chambers; the packed food-products will releases vapour pressure and bulging takes place, if counter pressure is not maintain the pouch ruptures. When processing is over with counter pressure; it enters the cooling chambers slowly cooling at different temperatures till it attains ambient temperatures. Comparing cost of packaging is quiet high and it has a very limited market in country like India. But Europe and U.S.A, it has a very good potential.

Radiation Processing in Food

More than thirty years of research and development at the Food Technology Division, BARC, has demonstrated that radiation processing of foods can contribute to nation's food security by reducing post-harvest losses caused by insect commercial potential for conservation of cereals, pulses and their products, spices, onions, potatoes, garlic, some tropical fruits, sea foods, meat and poultry.

Radiation processing can ensure hygienic quality in foods including frozen foods by eliminating food borne pathogens and parasitic organisms. It offers a viable environment friendly alternative t chemical fumigants and for quarantine treatment against insect pests in agricultural and horticultural products entering international trade. The safety and nutritional adequacy of irradiated foods for human consumption is well established. About 41 countries including India have regulations permitting irradiation of foods and 28 countries are irradiating foods for food processing industries and institutional catering.

Initiative on Commercialization

The first commercial demonstration plant for radiation processing of spices with an initial throughout of 30 tonnes / day is already operational in Navi Mumbai. This plant was set up by Board of Radiation and Isotope Technology (BRIT), a constituent unit of DAE. A commercial demonstration irradiator for the treatment of onions with a processing capacity of 10 tonnes / h is being set up by BARC in Lasalgaon, Nashik Dt., Maharashtra. Recently a NGO, Annapurna Mahila Mandal, has started selling radiation processed spices in the market in Mumbai.

Legislation and procedures for establishing a commercial food radiation processing facility

Regulations on Radiation Processing

Procedures for Approval of Radiation Processing of food

To oversee all aspects of commercial food irradiation in the country the Government of India have constituted a National Monitoring Agency under the Chairmanship of Secretary, Ministry of Health and Family Welfare.

Application for clearance for radiaton processing of specific food items are initially examined by the Expert Group on Food Irradiation constituted by NMA. The recommendations of this Expert Group are further considered by the Central Committee for Food Standards (CCFS) for approval. The food items approved for radiation processing are notified in the Gazette of India as an amendment to the Prevention of Food Adulteration Act, 1954 and Rules 1955.

Construction of Radiation Processing Plants

The construction and operation of a radiation processing facility for commercial purposes is governed by the Atomic Energy (Control of Irradiation of Food) Rules 1996, under the Atomic Energy Act, 1962.

Atomic Energy (Control of Irradiation of Food) Rules 1996 published in the Gazette of India, June, 1996 provide the Government statutory Rules for the authorization and licensing of radiation processing facilities for foods and procedures for obtaining certificate of approval from the competent authority (Atomic Energy Regulatory Board) and licence for a radiation facility from the licensing authority (Department of Atomic Energy). It also stipulates qualifications of personnel such as operators of radiation processing facility, radiological safety officer and quality control officer.

Clearances for Radiation Processed Foods

The Government of India has permitted radiation processing of the following food items for domestic marketing and consumption.

FOOD ITEM	RADIATION MINIMUM	DOSE (KGY) MAXIMUM	PURPOSE
Onions	0.03	0.09	Sprout inhibition
Potatoes	0.06	0.15	Sprout inhibition
Shallots(small onions) garlic, ginger	0.03	0.15	Sprout inhibition
Rice,semolina (sooji or rawa),atta(wheat flour) and maida(refined wheat flour)	0.25	1.0	Insect dinsinfestation
Pulses	0.25	1.0	Insect disinfestation
Dried sea foods	0.25	1.0	Insect disinfestation
Raisins, dried figs and dates	0.25	0.75	Insect disinfestation
Mango	0.25	0.75	Shelf-life extension and quarantine treatment for export trade
Meat and meat products including chicken	2.5	4.0	Shelf-life extension and pathogen control
Spices	6.0	14.0	Microbial decontamination
Fresh sea foods	1.0	3.0	Shelf-life extension
Frozen sea foods	4.0	6.0	Microbial pathogen control

Gray (GY=1 Joule/KG. 1 kilogray (KGY=1000 GY)

*Kilogray (KGY) in SI unit of energy absorbed by food from ionizing radiation

13

Shelf Life Studies

Introduction

The development phase of a new food product is progressing exactly according to schedule. Suddenly, an emergency status meeting is called. The product introduction must be moved up by six months. In the ensuing panic, the development schedule is scrutinized for possible shortcuts. Someone locates the category marked "Shelf Life" and asks, "Do we really need a full six months for this?" The answer is a resounding, "Maybe!"

For some food products, determining shelf life may only be a matter of days or weeks, barely affecting the timing of a product launch. For products with storage lives of months or years, the testing can add significantly to the development timetable. As the industry requires increasingly aggressive development cycles, the ability to perform valid accelerated shelf life testing gives a food designer a decided edge. A poorly conceived program can give disastrous results.

Shelf Life: Back to Basics

First of all, what is meant by shelf life? Many consumers believe that it refers to when a product becomes injurious to health or otherwise inedible. Some Food research Scientists, feel that a product reaches the end of its shelf life when "it no longer maintains the expected quality for the end user of a product." Each company has its own definition, which range from no acceptable changes to a predetermined degree of acceptable change in certain characteristics, including spoilage, flavor, texture, appearance and functionality.

Spoilage

For our purposes, spoilage refers to microbial or chemical processes which make a product unwholesome or even toxic. Products at risk for microbial storage provide an environment which supports microbial growth, and are not sterilized or are subject to contamination during storage. Chemical spoilage can occur through the leaching of chemicals or similar reactions promoted by long-term contact with packaging materials.

Flavor

Development of off-flavors during storage may be microbial in nature, as in fermentation, but often is the result of some chemical action. Oxidative reactions produce many off-flavors, including rancidity. Flavor degradation occurs with the volatilization of flavor compounds or with the absorption of foreign flavors.

Texture

Crackers and bread contain similar ingredients, but should not exhibit similar textures. As in most food products, water contributes greatly to the product texture. Other factors contribute to textural changes during storage including staling, breakdown of gel structures, phase separation, water activity, moisture migration and crystallization.

Appearance

Undesirable color changes including browning and fading can be caused by fat and / or moisture migration, chemical reactions, or be pH induced. Other processes, such as surface crystal formation, phase separation, syneresis and caking, produce an undesirable appearance during storage.

Functionality. If a vitamin fortified product loses its potency, or a chemical leavening agent loses its ability to produce gas, the product is no longer considered acceptable. Losing functionality during storage renders a product unusable.

A number of factors influence these product characteristics, many affecting more than one aspect including process and storage temperature; packaging barrier properties (to water and oxygen); atmosphere inside package (MAP, CAP); presence of light; presence and growth conditions of microorganisms; storage humidity; moisture and water activity; freeze / thaw cycles; chemical reactions (staling, crystallization); and pH.

These basics often are overlooked when setting up a shelf-life program, yet are critical to its success. According to some researchers, "It is essential that both the product and the packaging meet the specifications required or the results obtained from a shelf life study will not be valid."

The industry consensus is that to conduct an accelerated program, correlation with real-time data on a similar product under the same conditions must exist. A product which varies from the control study can result in inaccurate accelerated results.

Let's look at a hypothetical example. A company like Cadbury's or Nestle's, produces fig bars, sandwich cookies and low moisture chocolate chip cookies. All have been subjected to real-time versus accelerated shelf-life studies; time / condition relationships have been established. When the company has developed a high-moisture, fat-free chocolate chip and rice bran cookie. The program used for the original chocolate chip cookie would be inappropriate if applied to the new product because of gross product differences.

Start your Accelerator

To set up an accelerated program, list the product characteristics that define the product and the factors responsible for their deterioration. Set up a series of conditions to accelerate these factors. — Temperatures can be raised within certain limits above typical product storage conditions to accelerate reactions. Typical ranges are 20° to 32°F for frozen products; 45° to 50°F for refrigerated products; and 85° to 120°F for room temperature products. Keep in mind that higher temperatures are used for specific products to promote a certain reaction (*i.e.*, fat oxidation) but are not suitable if they promote other reactions at the same time.

A product subjected to cycling temperatures during normal storage must be subjected to the same cycling during an accelerated program. This proves critical when the food undergoes a change in phase, such as freeze/thaw cycles or the melting of fat in products with a solid fat phase, such as chocolate.

- Humidity can be used to accelerate product deterioration if the package allows the passage of moisture. The conditions used are dependent on both typical storage conditions and the moisture content of the food. Both high and low humidities may create deleterious effects if the product attributes are affected by moisture gain or loss.
- The effect of light on a product can be accelerated by extending the time or intensity of exposure. If the package is totally opaque, testing the effect of light is unnecessary unless considering the deterioration of the packaging material itself.
- Accelerating oxidative reactions by increasing oxygen pressure has met with some success, but the high temperatures required limits its usage.

Environmental rooms that control temperature, humidity and light levels can be made or purchased so that the test material is not subjected to unplanned fluctuations. Products should be held under multiple accelerated conditions in addition to real-time conditions, to obtain a more accurate picture of the factors at work.

The amount of product needed for testing is determined by the number of tests, their frequency and the number of conditions used. Increase this amount to provide for additional testing in case of some inexplicable test result. All products must be made under the same conditions to avoid possible differences in process variables, ingredient loss, etc., that may affect the product shelf life. The specific tests performed will depend on the characteristics defining the end of shelf life. Test frequency depends on the severity of the accelerated conditions. More severe conditions require shorter test intervals to obtain meaningful data points and to determine the actual end point. Testing includes sensory, chemical and physical analyses. The method of sensory testing should provide a significant difference between the product undergoing testing and fresh, equilibrated "control" product.

"Don't be afraid to use a numerical control based on the initial scores of the product. The numbers may be more valid than those obtained from new or improperly stored control product," is what one of the scientist say.

Chemical and physical tests should quantify the attribute showing deterioration. These can run the gamut from peroxide value or free fatty acid analysis (FFA) for rancidity, to percent water and water activity (Aw) for moisture changes; compression tests for textural changes; and volume tests for whipping ability.

When possible, correlate chemical, physical and sensory data. This can be fairly straightforward in the case of texture measurements versus moisture, or as complicated as in the case of color acceptability versus a reading on a colorimeter. A linear relationship may not exist. A low acceptance score may be related to both a high and low physical reading if the optimum is in the medium range, such as in the case of a product that can be both too soft and too hard. After the product reaches a predetermined end point, data between the various accelerated and real-time conditions can be analyzed and relationships determined. Mathematical models to judge the rate of reactions for accelerated conditions can then be established. Be forewarned that different degenerative reactions may proceed at different rates under the same conditions in an accelerated test. Textural unacceptability, for example, may occur before rancidity in real-time storage, but the opposite may happen in an accelerated study.

What else can go wrong?

Accelerated shelf-life testing has other limitations. One of the reasons for this is that the extreme conditions used to accelerate degeneration may create effects that would never occur during normal storage, such as protein denaturation at high temperatures; an increase in Aw at high temperatures affecting reaction rates; concentration of reactants in solution during freezing changing reaction rates or resulting in precipitation; transfer of reactants via migration in melted fat; decrease in oxidative reactions due to decreased oxygen solubility; and inhibition of pH-dependent reactions due to temperature dependent solubility.

Another factor to keep in mind is that "accelerated" conditions may be encountered during routine storage. Going back to our friends at the Wildly Successful Cookie Co., a study has just established that a new product reaches its end point after a six week storage period at 100°F / 80 per cent relative humidity (RH). WSCC designers conclude that this correlates to one year at ambient conditions of 68°F / 35 per cent RH. What this actually means is that the product has a shelf life of only six weeks when stored without air conditioning in Florida during the summer. A final hurdle is that the barrier properties of packaging may change during accelerated conditions. The permeability of film to water may change with temperature.

The Alternatives

Not very many alternatives exist to replace shelf-life studies. Applicable data for a specific product may be found with a search of technical literature. Because of the complexity of most food products, finding direct correlations is not likely. If the product has an end point directly related to moisture or water activity, one can predict shelf life due to moisture changes using the water vapor transmission rate (WTVR) of the packaging material. The critical Aw or moisture content must first be determined. While this determination still can be a challenge, newer computer programs can take formulation data and produce a corresponding water activity and predict a theoretical shelf life. The same type of calculations can be made in regard to fat oxidation. The rate of oxidation of a food system is a function of oxygen partial pressure which is affected by the permeability of the packaging material to oxygen. In short, the ability to predict product shelf life is a complex proposition. The more familiar a food designer is with the factors that affect the system in question and the more extensive the previous testing, the more likely that "maybe" can change to "We can have a high degree of confidence in our accelerated program." Simply the time it takes for a food product to become unacceptable the time during which the food will:

- Remain safe
- Retain desired sensory, chemical, physical and microbiological characteristics
- Comply with label declaration of nutritional data

Definition does not mention storage conditions which can vary widely - room temperature caries, refrigeration temps vary, freezer temps vary Designated by "Use by", "Best before" date information. Determined by combination of production, storage, distribution and consumer handling aspects.

Factors Affecting Shelf Life

1. Intrinsic

- Water activity
- pH/acidity
- Redox potential
- Oxygen availability
- Nutrients
- Natural/surviving microflora
- Biochemistry of product (enzymes, chemical reactants)
- Preservatives

2. Extrinsic

- Time-temperature of processing
- Temperature of storage
- Relative humidity

- Exposure to light
- Microbial contamination during processing, packaging, storage
- Atmosphere composition
- Heat treatment before consumption
- Consumer handling
    - Short shelf-life products - microbial growth most important
    - Long-shelf life products - non-microbial factors most important

Determination of Shelf Life

Done by storage trial

Should be done on:

- First reasonable prototype (at this stage final packaging may not be possible)
- Pilot plant runs
- Full-scale production runs (preferably unsupervised by development personnel). Shelf-life trials should be done on product from three different runs

For shelf-life trials need to:

1. Determine the necessary tests - microbiological, chemical, physical, sensory
2. Set acceptable limits for test results, *e.g.* total count of 107 cfu/ml
3. Develop a testing plan based on desired shelf life
    - numbers of samples
    - frequency of testing
    - use good sample for comparison (freeze fresh sample, prepare "standard" sample) - depends on the product
    - maximum time of test
    - temperature of storage, often more than one temp used
    - normal temperature of storage
    - 'abuse' temperature
    - 'high' temperature in accelerated trial *see below*
    - fluctuating temperature to simulate changes during transport, etc
    - other conditions of storage, eg
    - relative humidity for samples likely to absorb or lose moisture
    - physical movement which may cause crushing or breaking (can use mechanical shaker or place shelf-life samples in moving vehicle)
    - light exposure *e.g.* fluorescent lights

Shelf-life for product labelling purposes, *e.g.* 'use-by date', set with margin of error, when generally 70 per cent of time to spoilage as determined in the shelf-life trials

Accelerated Shelf-Life Testing

Performed to save time - shelf-life trials can often delay launch of new products Can shorten trial but product may not spoil / deteriorate in same manner as at normal temperatures, *e.g.* staling of bread is slower at higher temperature

Other changes can occur at elevated temperatures which can cause under- or overestimation of shelf-life:

- phase changes, *e.g.* melting of fats
- crystallisation changes, *e.g.* cocoa butter, starches
- change in relative rates of reactions
- increased water activity
- denaturation of proteins
- decreased solubility of gases

Therefore need to test validity of using accelerated conditions against changes which occur at normal temperature, but over a longer time

Still need to know how to interpret results from accelerated trial. *e.g.* if a product's shelf-life is 3 weeks at 45°C, what is it at room temperature?

	Days of storage								
	0	1	2	3	4	7	14	21	28
Coliforms	*								
Yeasts and moulds	*					*	*	*	*
Gas production at 7°C				*	*	*	*	*	*
Gas production at 21°C		*	*	*	*	*	*	*	*
Texture & syneresis		*				*	*	*	*
pH/acidity	*	*				*	*	*	*
Flavour		*				*	*	*	*

Storage temperature 7°C unless otherwise stated

Target shelf life: 21 days

* sampling/testing times

Illustration of Shelf-Life Testing Schedule for a Yoghurt

Food Preservation Techniques in a Nutshell

CANNING first destroys bacteria through heating and then the food is placed in a sterilized container and sealed.

DRYING removes water from the food that is required by spoilage bacteria to grow and reproduce.

FREEZING slows down the spoilage process by changing that some essential water into ice, a form that the bacteria cannot use.

PASTEURIZATION destroys most of the existing spoilage organisms by heating the food to a high temperature for a short duration.

PICKLING OR FERMENTATION (culturing) leaves the food with a higher level of acid, making it an inhospitable environment for spoilage bacteria.

VACUUM PACKAGING uses a vacuum sealed, abrasion-resistant moisture-impermeable film that inhibits molds, yeasts, and bacterial growth on the surface of the things such as meat. Since there is no air in the package, vacuum-packaged meat will have a darker, purple color before being opened. Once the meat is exposed to oxygen, it will turn the familiar bright red color, because of the natural reactions within the package. Fresh vacuum packaged meat will give off a slight odor upon opening. The smell will dissipate within a few minutes. This should not be confused with spoilage.

SMOKING adds smoke-born chemicals to food that help destroy potential spoilage organisms.

CHEMICAL ADDITIVES are designed to destroy spoilage organisms or inhibit their growth.

Sugar and salt are examples of additives that have been in use for centuries. Both of these work by drawing water out of the spoilage organisms, thus preventing their growth.

UHT - ultra-high temperature, higher than pasteurization resulting in a sterile product.

IRRADIATION - process like pasteurization that pasteurizes food by using energy, just like milk is pasteurized using heat. Irradiation DOES NOT make food radioactive. The food never touches a radioactive substance. Irradiation destroys insects, fungi, and bacteria. Fewer nutrients are lost during irradiation than in cooking and freezing. Food irradiation has been approved in 37 countries for more than 40 products. Astronauts have eaten irradiated foods for years.

FOOD ADDITIVES - a food additive is any substance added to food. Sugar, salt, and corn syrup are the most commonly used food additives. Food additives keep foods fresh, slow microbial growth, give desired texture and appearance, aid in processing and preparation.

14
Sensory Evaluation

Introduction

SENSORY EVALUATION is defined as a scientific discipline used to measure, analyze and interpret reactions to those characteristics of materials as they are perceived by the senses of sight, smell, taste, touch and hearing.

Sensory Evaluation by Taste Panels

The taste panel is selected based on the following consideration

a) Function of a panel within the total sensory evaluation system
b) The types of the panels and there functions – under this it should be understood that how panel work fit into overall sensory evaluation program, choosing qualified and motivated persons, thorough training in terminology by using a broad and comprehensive frame of references for each product, introducing a scaling method which can be taught effectively and which can be treated statistically and using the panel data effectively to study products in depth and provide detailed product analysis to management.

Various Methods Used for Sensory Evaluations by Taste Panels

See figure on Page No. 138.

Discrimination

It determines how products perceived and and significance of difference. it requires trained panel example – triangle test, paired comparison

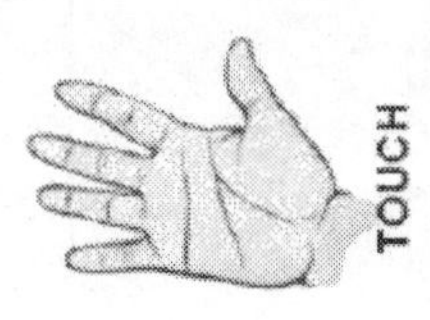

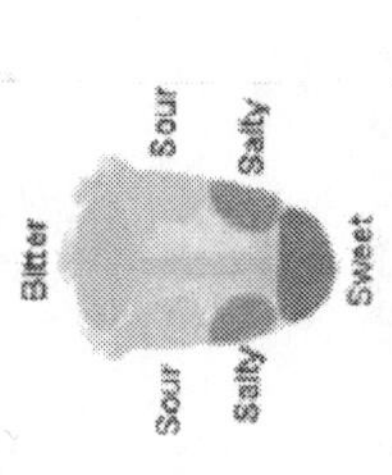

SIGHT
Vision = 0.02s
SMELL
Odour = 0.002s
TASTE
Salt = 0.3s ,
Sweet = 0.4s,
Sour = 0.5s,
Bitter = 1.0s
Bitter
Sour
Salty
Sweet
Sour
Salty
AUDITION
Hearing = 0.01s
TOUCH
Touch = 0.005s

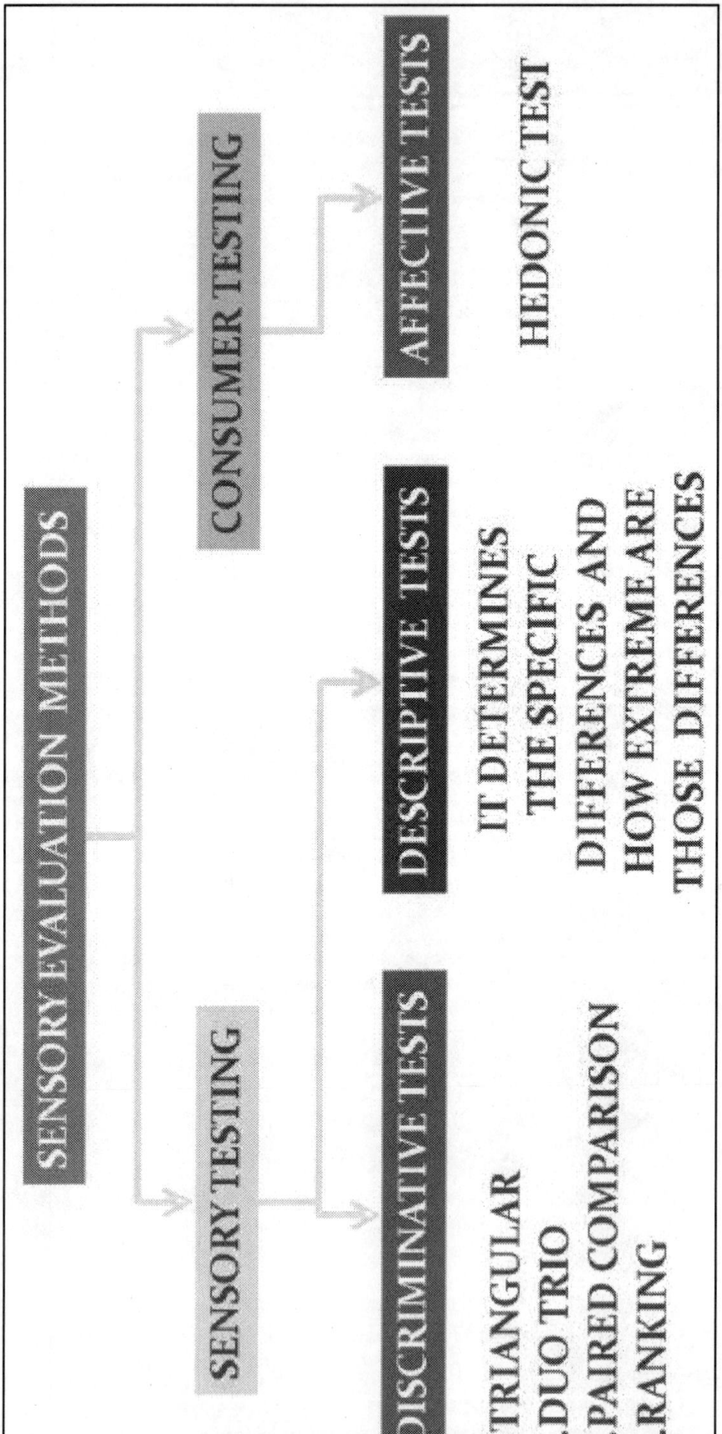

SENSORY EVALUATION METHODS

SENSORY TESTING

CONSUMER TESTING

DISCRIMINATIVE TESTS

1. TRIANGULAR
2. DUO TRIO
3. PAIRED COMPARISON
4. RANKING

DESCRIPTIVE TESTS

IT DETERMINES THE SPECIFIC DIFFERENCES AND HOW EXTREME ARE THOSE DIFFERENCES

AFFECTIVE TESTS

HEDONIC TEST

Descriptive

It determines the specific differences and how extreme are those differences. It requires trained panels example – Linear scorecard.

Affective

It measures preferences for products or magnitude of like / dislike for a product. It can be used for consumers or trained panelists example Hedonic test.

Discrimination

Triangle Test

Panelists receive three coded samples. They are told that two of the samples are the same and that one is different. Panelists are asked to identify the odd sample. This discriminatory testing method is often used as a tool in quality assurance programs to insure that samples from different production lots are against yours. Triangle tests are also used in product development studies to determine if various ingredient substitutions or changes in processes will result in adverse product effects.

Duo-Trio Test

In this test, three samples are presented: one sample is labeled "R"(reference) and the other two are coded. One of the coded samples is identical to "R" and the other coded sample is "R". Panelists are asked to identify the correct sample by choice is always 50 per cent. Both Triangle and Duo-Trio may be used to screen panelists for their ability to repeatably select a specific trait when tasting products for flavor, the Duo-Trio test is often used instead of the Triangle test because the former method requires less tasting.

Paired-Comparison Test

A pair of coded samples is presented for comparison on the basis of some specified characteristic (saltiness, sweetness etc.) This method is similar to the triangle tests, however fewer samples are required and there is less tatting. The statistical efficiency is not, however, as great with the paired- comparison as it is with triangle test.

Ranking Test

This test is an extension of Paired- comparison test. Panelists receive three or more coded samples are asked to rank samples for intensity of some specific characteristic. Ranking tests are often used to screen one or two "best samples' from a group of samples rather than to thoroughly test all samples. However, no indication of the magnitude of difference between samples is obtained because samples are only evaluated in relationship to each other.

Descriptive Methods

It involves discrimination of qualitative and quantitative traits of a sample by use of a small (5- 10) member trained panel. Such techniques as flavor profile, textural profile and QDA (quantitative descriptive analysis are used. Descriptive analysis requires at least three evaluative processes. First, discrimination of the trait ; second, description of the trait; and third, quantifying the trait. The steps of discrimination and description of traits are qualitative. The language used is developed through careful training and practice with the panel. This requires panelists to develop a common vocabulary which catalogues various sensory stimuli with appropriate language. Descriptive analysis is a complex cognitive process which requires more mental acuity than sharp taste or olfactory senses. Discrimination among stimuli is only part of the qualitative process. The third step in descriptive analysis is to quantify the traits as to how strong they are. Two products may be quite similar in their qualitative components but differ overall because of the relative intensities of these characteristics. Descriptive analysis is the only sensory method that deals with the total picture or profile of a food product. it may be used to suggest or interpret instrumental methods or other sensory test information or be used as a research guideline tool, or as a method of quality assurance or quality control.

Scale Tests

Many types of scales or scoring systems are used for preference evaluations. These may include structured word scales or graphics. In structured scales a scale of 7 to 10 points is recommended because panelists tend to avoid using the end points on the scale. To use fewer than 7 point scale may not allow the panelists to show the degree of variation observed. More of a word-descriptive scale becomes the most critical feature. It may be appropriate to involve the panel n defining word descriptions and numerical scores to be assigned to these words. Remember, the panel results must be interpretable and accurately reflect what was subjectively measured.

Affective Test

Hedonic Scale

The word "hedonic" is of Greek origin and relates to degree or magnitudes of like or dislike.

Rank Preference

When more than 4 or 5 samples are served in a ranking test, the difficulty for the panelists to rank products is increased. More re-testing is generally done in order to assure the correct positioning of the rankings.

Paired Preference Test

This is the most simple and oldest method for to the panel. The panel is asked to indicate which sample they preferred and why they chose the preferred sample. In all preference testing at least one sample (control) with a known hedonic value (scale of like or dislike) should be included. Samples may be ranked from most to least desirable even when all samples are considered to be of questionable desirability. A known sample provides a reference point upon which to base results.

15

Packaging of Specific Food

Preservation of Fruits and Vegetables

The magnitude of post harvest losses in fresh fruits and vegetables is estimated to be 25 – 80 per cent depending upon the commodity and the technological level of post harvest operations. This reflects the lack of knowledge by post harvest handlers of the biological and environmental factors involved in the deterioration or the absence of adequate post harvest technologies required preserving fresh quality. In recent years, much attention has been paid to explore the potential of surface coating to maintain the quality of harvested fresh produce and to reduce the volume of disposable non-biodegradable packaging materials. Many fruits develop a waxy coat on their epidermis as they mature on the plant (apples, bananas, mangoes, and grapes) they develop a waxy coating on the epidermis, which is either felt or seen as a powdery bloom. Wax usually develops when the fruit has attained 2/3 of its growth. But much extraneous matter and spray residues. The natural waxy coat is not adequate to offer protection against water loss and high respiration rate. Thus the approach towards prolonging the life of fruit/vegetable involves restricting the rate of respiration and preventing moisture loss. So as to maintain the vital food elements in as near the same quality as in the freshly picked fruit or vegetable. The growing importance of fruits and vegetables in commerce has led to development in various protective wax coatings with and without fungicides, bactericides, growth regulators, etc., wax coating of fresh fruit and vegetable is quite old and it has great potential in the storage and transportation of fresh produce.

Surface coatings have been used extensively on bulky organs to modify internal atmosphere composition and thereby delay ripening, reduce the water loss, and it improves the finish of the skin. Post harvest losses can be reduced to some extent by, selection of varieties, careful handling, cleaning and sorting, control of ripening process, control of spoilage, packaging, use edible films, use of fungicide, use of ethylene absorbent, use of sprout inhibitors, use of growth regulators, irradiation, packaging home operations, dumping, washing and waxing.

Waxing

Waxing of fruits and vegetables is done according to the recommendations. Food grade waxes are used to replace some of the natural; waxes removed in washing and cleaning operations, and this helps to reduce the water loss during handling and marketing. If produce is waxed, the wax coating is allowed to dry thoroughly before further handling. Waxing of the produce is an old age art, which was started in the beginning of the 19th century. Waxing was used as a preservation technique for fruits since 1900. Wax has been used since pre historic times. In 4200 BC, the ancient Egyptians kept bees and used the wax from honeycombs to make models and to preserve their dead and in encaustic painting. The Romans used in pottery making textile printing, as early as 12th century, the Chinese were using waxing processes on fruits, but it was not until 1922 that a waxing process was introduces for wide spread use on fruits and vegetables.

Wax

Wax is an ester a long chain aliphatic acid chain aliphatic alcohol. Waxes are the esters formed from a fatty acid and a high molecular weight alcohol.

Principle Involved During Storage

The produce continues respiring and uses up all the O2 within the fruit which is not being replaced as quickly as before waxing because of the coating and produces CO2, which accumulates within the fruit because it cannot escape as easily through the coating.

Eventually the fruit will switch to anaerobic respiration that does not require O2. An aerobic respiration produces much more CO2, acetaldehyde and ethanol. The acetaldehyde and ethanol give off – flavour to the product, which are detrimental to the perceived quality. The natural barrier of the fruit and vegetable, the type of wax and amount if was applied will influence the extent to which the internal atmosphere (O2 and CO2) are modified and the level of reduction in weight loss. All fruits and vegetable are living organisms and as such require intake of O2 and release CO2 for respiration (aerobic). They also release vapour top the atmosphere through diffusion. Both these processes create a loss in weight of the produce. All fruit and vegetables are covered naturally in a cuticle, which is barrier to gas exchange. The cuticle is very impermeable to O2, slightly more permeable to CO2 and several magnitudes more permeable to water vapour. Unwaxed fruit or vegetables have pores in their surface through which most the diffusion of O2 and CO2 offers. Water vapour can move through the pores, cuticle and micro–cracks in the cuticle. During the process of waxing a tightly adhering thin film of the coating substance is applied to the surface of the fruit. The surface coating is often relatively impermeable to O2 and CO2 and water.

Protective Coating on Fresh Produce

Human became more health conscious. This gave new awareness of fresh fruits and vegetables, and resulted in rapidly increasing demand. Consumption of fresh produce, to help the consumers to obtain possible product, Fruits and vegetables are living organisms, which contain 80 - 90 per cent water by weight. If they are left without cuticle, the water quickly begins to evaporate, resulting in poor product shelf life to increase freshness and appearance. The fruit and vegetable growers should concern about consumer safety. There are few active ingredients which when combined with water, wetting and drying agents and form the compounds used a protective coating. The coating may contain one or more of the following. Carnauba wax, shellac wax, paraffin wax, candelilla wax, bees wax.

Different Types of Waxes

Animal Waxes

- Bees wax.
- Spermaceti wax
- Shellac wax
- Chinese insect wax

Vegetable Wax

- Carnauba wax
- Candelilla wax
- Sugarcane wax
- Palm wax
- Esparto wax
- Japan wax
- Oricury wax
- Waxol 0.12, is the vegetable wax manufactured by Central Food Technological research Institute (CFTRI), Mysore.

Mineral and Synthetic Wax

- Ozocerite
- Montan wax
- Synthetic wax

Foodstuffs that are waxed are:

Fruits

Apple, Avocadoes, Bell pepper Lemon Grapes, Banana Melons, Oranges Lime Passion fruit, Peaches Pine apples,

Vegetables

Cucumber Tomato Sweet potato Melon, Methods of waxing:

Four principle methods of waxing of fruits and vegetables are there,

1. Liquid Paraffin Wax Method

In this method fruits and vegetables are dipped in hot paraffin. Some times resins are added. The main disadvantage of this method is too much of coating material is used.

2. Slab Wax Method

In this case the wax is pressed against rapidly revolving brushes. But the efficiency is very less.

3. Spray Method

Spraying of melted wax on the fruit, which is subsequently brushed mechanically until a film of desired thickness is obtained. The wax is dissolved in a suitable solvent. This depends on,

- The pressure employed
- Volume of wax used
- Wax temperature
- Distance of fruit from the spray
- Number of spray nozzles.

4. Dipping or Cold Wax Method

Fruits and vegetables are washed and then without being dried are dipped into a wax emulsion of proper concentration. They are dried before packing. Purified wax is odour less, tasteless and nontoxic and it can be heat- sealed.

Properties of Wax

- Kneadable at 20°C.
- Easily emulsifiable.
- Should not impart undesirable odour.
- Should be economical.
- Efficient drying performance.
- Non-sticky or tacky.
- Should never interfere with the quality of fresh fruit / vegetable.
- Melts above 40°C without decomposition.
- Has relatively low viscosity.
- Capable of being polished by slight pressure.
- Translucent to opaque form but not like glass.

The properties of wax depend primarily on molecular structure rather than molecular size and chemical constitution.

Purpose of Waxing

- To reduce the weight loss.
- To replace the natural wax.
- To increase the appearance
- To increase the freshness
- To decrease the rate of transpiration
- To reduce the shrinkage losses.
- To inhibit mold growth.
- To prevent other physical damage and disease. Necessary steps involved in waxing
- There should not be any production of off-flavours.
- Rapid drying of the product after waxing on the fruit.
- Manufactured from good grade materials.
- Competitive price.

During harvesting, transport, washing, grading, it can undergo some scratching or abrasion, which removes natural layer or was and also scars the protective layer of skin.

Lipids and Resins

Lipids and resin are added to coating formulations to reduce gas exchange, but mainly to impart hydrophobicity (to reduce water loss) and gloss appearance. Fruits coated with resins will develop whitening of the skin due to condensation that develops when they are brought from cold storage to ambient conditions.

Composite and Bilayer

Several composite and bilayer films have been investigated with goal of combining the desirable properties of different material to improve permeability characteristics, gloss, strength, flexibility, nutritional value, and general performance or coating formulations.

Plasticizers have been incorporated into edible coatings as a processing aid to facilitate coating application and to increase merchantability. The most commonly used plasticizers are polyols (sorbitol and glycerol, mono, di or oligo saccharides, lipids0. the plasticizing effect is due to their ability to reduce internal hydrogen bonding while increase intermolecular spacing. This will reduce the brittleness of the film but increases the permeability of water. Several wax formulations have been developed and tasted experimentally. Suitable emulsifiers are used to produce a wax emulsion. Synthetic or natural resins are added to the wax emulsions to give more gloss to the treated produce. Water-wax emulsions are safer to use than solvent waxes, which are highly flammable. Water -wax emulsion can be used without drying the fruit prior to the treatment, therefore washing and wax treatment can be combined. Carnauba wax, sugarcane wax, thermoplastic terpene resins and shellac are common substances, which are used commercially; however waxes of plant origin are preferred. Development in wax emulsions were not limited to the formulation of new types of wax, but also included the addition of chemical fungicide, which gave waxing treatment an additional function. For example, biphenyl amine added to the wax emulsion not only gave additional shine and reduced weight loss and respiration rate of apples and pears, but also reduced the occurrence of scald. Addition of post harvest fungicides to the wax, emulsion to.control Penicillium rot in stored citrus fruits also proved helpful. Waxes may be applied by dipping, spraying, foaming or brushing. Foaming is a better method of application since it leaves a very thin layer of wax on the fruit. If a thicker layer of wax is applied to the fruit surface, it becomes an undesirable barrier between the external and internal atmosphere and restricts exchange of respiratory gases (CO2 and O2). This may result in anaerobic respiration, resulting in fermentation and development of an off flavor. It is therefore necessary to adjust the thickness of the wax coat according to the variety and storage and marketing temperatures.

Research on the Effect of Wax Coatings

The effect of the wax emulsions Fruitex, Britex-561 and SB65 on oranges,Kinnow, lemons and grape fruits was studied during refrigerated and non-refrigerated storage (Farooqi *et al.*, 1988). Wax Coating of fruit improved its external appearance by giving additional shine to the fruit surface. It reduced weight loss and kept the fruit firmer, thus marinating its fresh look, while it reduced rates of respiration and ethylene production, thus delaying senescence. No significant change in constituents like Vitamin C, acidity, and sugar contents (In citrus fruit especially) were recorded due to treatment. However higher concentration of wax resulted in the development of off flavor in citrus fruits particularly during holding at 20° C, making this treatment unsuitable for this fruit. It was, however found necessary to tailor the thickness of the wax coat according to storage temperature to avoid development of off flavor. FDA regulates these waxes, or coatings, as food additives approved or "generally recognized as safe" for human consumption. However, some consumers have concerns about their use. Vegetarians and others who avoid animal products may worry that fruits and vegetables contain animal-based waxes, such as oleic acid.

Some people fear that the wax traps pesticides, making the fruit or vegetable unsafe to eat— even though FDA's pesticide monitoring program indicates that pesticide residues on fruits and vegetables are consistently within acceptable safe limits. If consumers want to avoid waxed fruits and vegetables, FDA regulations that took effect in 1994 may help them identify the appropriate products for them. These regulations require produce packers or grocers to provide point-of-sale information about the presence of waxes on fresh fruits and vegetables. This information can appear on labels of individual products, packing cartons (if they are used at the point of sale), or on counter cards or signs. The information will say that the product is:

- Coated with food-grade animal-based wax to maintain freshness, or

- Coated with food-grade vegetable-, petroleum-, beeswax-, and / or shellac-based wax or resin to maintain freshness. If only one of these types of waxes is applied, the label can simply identify the type, such as "vegetable-based." FDA also will allow the statement "No wax or resin coating" on fresh fruits and vegetables that do not contain wax. Besides reading labeling information, consumers can reduce their concerns about waxes by rinsing fruits and vegetables with warm water and, when appropriate, scrubbing with a brush. This will eliminate much of the wax.

New Developments in Coating

Scientists are hard at work developing new methods of coating food to help it keep its freshness and visual appeal even longer—so a refrigerated slice of apple, for instance, could retain its fresh-cut look for up to three weeks, instead of turning brown in short order. Researchers at Food Science Australia, for example, are looking at creating more coatings made from naturally occurring compounds found in the foods themselves. Called modified atmosphere coatings, or MACs, these new coatings would allow only certain gases to pass through, preventing the food's deterioration without the need for plastic packaging. At the University of Georgia, food science and technology professor Manjeet Chinnan is studying both existing and new coatings and "edible films"—wrapping material that is made from plants instead of plastic. Edible films—based on the plant-based wrappings people in Asia have been using for centuries for various foods, such as sticky candies— being either eaten or at the very least biodegradable, could greatly reduce waste and the use of natural resources, since they're often made from food-processing byproducts. They have found limited use so far, but further research could lead to their gradually supplanting the ubiquitous plastic film almost everything is wrapped in today. Chinnan has discovered coatings that can increase the shelf life of foods like roasted peanuts by up to 60 percent and plant-based coatings that reduce the fat absorption of fried chicken strips by up to a third (and the chicken remained moister, to boot). Other coatings he's studied prevent the fats and acids in food being fried (vinegar and lemon juice in a marinade, for instance) from migrating into the frying oil, which keeps the frying oil fresher and allows it to be used longer. In future, Chinnan's team hopes to engineer films that could transfer flavors to foods, add vitamins, or even contain anti-microbial agents that could reduce the risk of microbial contamination, and hence the risk of food poisoning. For something food coatings accomplish a lot.

Conclusion

Coating improves product appearance, colour, crispness, flavour, nutritive value, juiciness, texture, etc. From fruits / vegetable to meat products are can be coated including chocolates, various nuts, wafers, French fries, chicken, fish, corn chips, apple, lemon. Though coating is an effective method of value addition and preservation for the food product, it is having some problems. Continuous Mechanization is not available for various coating operations. Till now, many of the coating operations are done by batch processing. Safety of food product should be maintained in the coating operations. For example reuse of batter in battering operation will induce the chance of contamination over the food product. Many of the coating materials are high in cost and some of the coating operations are also higher cost. This leads to increase the cost of operation of coating. So, we have to reduce the cost of operation by finding out the alternative coating martial and method instead of costly coating material and method.

Packaging of Dairy Products

What are Dairy Products?

Dairy products are generally defined as foods produced from Cow's or Buffalo's Milk. They are usually high-energy-yielding food products. A production plant for such processing is called a dairy or a dairy factory. Raw milk for processing mostly comes from Cow's Milk or Buffalo's Milk. Milk is a liquid and therefore requires a container at every stage of movement from the cow to the consumer. At the early stages of dairy development the cow's udder was used as the basic container for all purposes. The cow, kept in the town cow shed (*tabela*), was brought to the customer's doorstep for milking. In some cases the milk was sold from a shop adjacent to the cowshed. In many European countries town cow-keepers could still be found after the First World War but, for reasons of hygiene and economy, they quickly disappeared. This trend seems to be unavoidable for the dairy industry worldwide and will certainly be applied to cities in developing countries where town cow-keeping still exists. The growing demand for milk in towns and the high costs of milk production within their boundaries led to the development - probably around 1860–70 - of containers suitable for various stages of marketing and distribution. These were metal cans, provided with a lid and having capacities up to about 80 litres. The introduction of this type of container (until recent years often called a 'churn') facilitated the transport by railway from rural areas to towns, thus contributing substantially to the rapid growth of milk distribution. Similar containers were also

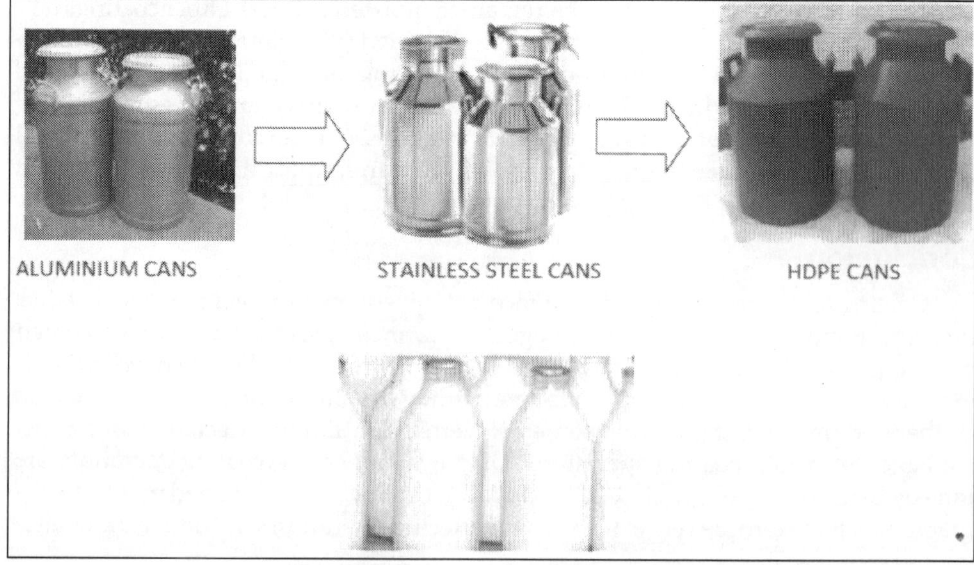

used for retail delivery to the consumer, the milk being dispensed in the street or at the doorstep into the consumer's container.

The first significant development in the packaging of milk for **retail sale** came at the very end of last century with the introduction of the process for sterilized milk in which the retail container, the glass bottle, formed an integral and essential part. In 1930 bottling of pasteurized milk developed rapidly, first in America and soon after in Europe.

The glass bottle as the retail package for milk remained unchallenged until 1933 when the first carton made of waxed paper was introduced. The development and introduction of plastic materials for packaging in the dairy industry (initially polyethylene in 1940), alone and in combination with paper, resulted in a wide range of containers, termed cartons, suitable for liquid milk. When we refer to liquid milk we usually mean a product, either processed or, less often, sold raw to the consumer, deriving from a lactating ruminant, mainly the cow. Processing depends on the grade of milk to be manufactured following the regulations and customs of the country. Heat treatment and, in most countries, standardization of butterfat content, are the basic parts of the processing procedures.

Heat Treatment of Milk

Heat treatment may be classified as:

- **Pasteurization**
- **Sterilization (In Bottle)**
- **UHT (Ultra-High-Temperature) Treatment Integrated With Aseptic Packing.**
- **Pasteurization HTST** ***(high-temperature short-time - heating at 72°C for 15 seconds)*** fulfils the following main objectives:
  - to safeguard public health by destroying all pathogenic bacteria
  - to extend the keeping quality of liquid milk by destroying most of the milk-souring micro-organisms
  - to ensure a product with a good keeping quality.

Sterilization (in bottle) is the term applied to a heat treatment process which has a bactericidal effect greater than pasteurization. Although it does not result in sterility, it gives the processed milk a longer shelf life. This is achieved partly by using a more severe heat treatment ***(about 110°C for 20–30 min)*** and partly by applying the treatment after the bottle is filled and sealed which eliminates the risk of contamination during packaging. As a result of the long holding time at this elevated temperature, the product has a cooked flavour and a pronounced brown colour.

UHT treatment is a process of high bactericidal effect, developed as a continuous flow process in which the milk is heated ***at 135°C–150°C for about two seconds only***. This treatment must be integrated with aseptic packaging in sterile containers. UHT milk has less pronounced cooked flavour and no brown colour. As a criterion for packaging requirements for pasteurized milk in general, a shelf life of several days at a temperature below 10°C can be assumed. In bottle sterilized milk can normally be kept for weeks and UHT milk aseptically packaged can be kept for several months, both without refrigeration, provided the package is not opened. After opening, the sterility of the product is lost and the shelf life becomes close to that of pasteurized milk. Sales of unpasteurized milk are rare in countries with a developed dairy industry and often prohibited by law. Nevertheless, there is some where purchased milk is boiled at home as a common habit even though the raw milk is of a high hygienic standard. In such cases heat treatment in the milk plant may be considered as an unnecessary expense and not required by law.

Packaging of Milk for Retial Sale

Packaging of milk for retail sale can be two ways.

1. Returnable Pack
2. Single Service Pack

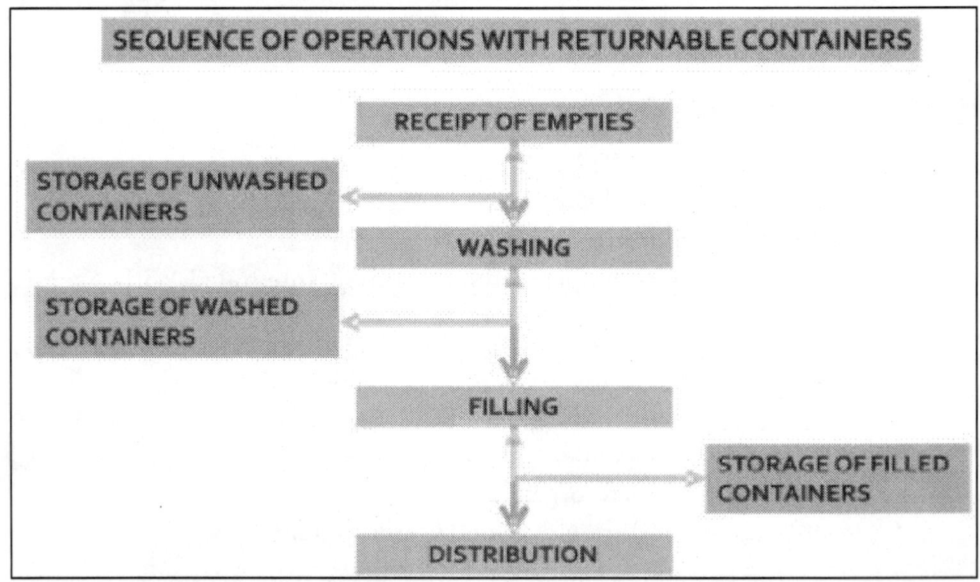

Returnable Pack

The basic features of systems using returnable containers are the collection of empties and washing prior to re-filling. Differences in operation times and capacities of the various machines involved make intermediate storage necessary. Storage of unwashed empties is normally essential and may extend overnight so that washing and filling operations can begin next morning before the day's supply of unwashed empties arrives. Storage of washed cans is permissible as they have lids but storage of washed bottles is extremely bad practice because they are unsealed and therefore liable to contamination. Normally storage must be provided for filled cans and bottles to give flexibility in the distribution arrangements. For pasteurized milk this must be refrigerated.

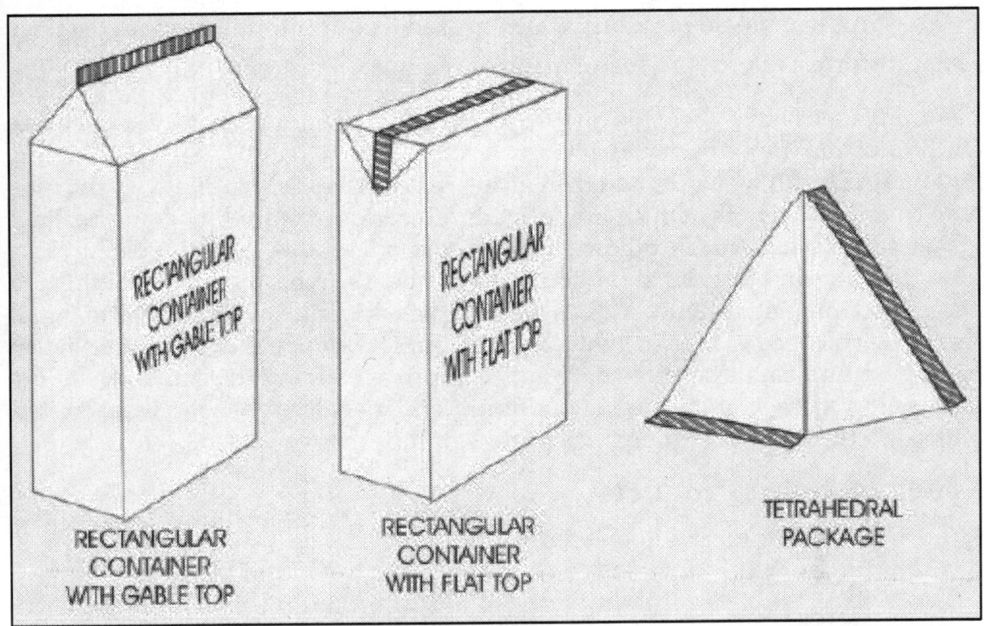

Bottle washing, filling and capping machines should be of matching capacity, otherwise the labour-intensive operations of decrating and crating, as well as unstacking and stacking, would have to be repeated unnecessarily. This problem does not arise with cans, since they are not crated and may be easily stored empty after cleaning. The required storage area, both for empties and for product, depends on the operation schedule of the plant which in turn is affected by the relation between required dispatch capacity and the capacity of filling machines in operation. As its is evident, the required storage areas may differ considerably depending on the arrangement of working time, type of packaging and capacity of equipment. Availability of space and monitoring the stock is a major job.

Single Service Pack

The common feature of single-service containers is that after emptying they are discarded. This fact has a significant impact on the milk plant construction, organization, and on the economics of the whole enterprise. There is no collection and washing of the milk packages - only crates are collected and washed, but even these may be replaced by single-service delivery wraps, trays or boxes. Palletization may be applied as in the case of returnable containers. Intermediate storage of packing material and filled packages is required and this must be provided in the plant. Two basic types of single-service containers are considered in this analysis, *i.e.* cartons and plastic sachets. Cartons are usually made in one of the shapes. The material used in each case is polyethylene-coated paper-board; in aseptic versions an aluminium foil lamina is normally incorporated. There are three methods in which cartons can be made.

- Pre-formed cartons

- pre-cut, single-piece blanks pre-creased ready for forming into a container
- form a tube from a reel of material and make a carton

Container with a gable top and these generally are more expensive. The other method of making a carton is to form a tube from a reel of material, seam it longitudinally, fill with milk and then make transverse seals. Alternatively the tube may be cut into lengths which are formed into cartons before filling and sealing. Plastic sachets are usually pillow-shaped and made of low density polyethylene film. They may be reeled single or double film or lay-flat tube, the latter avoiding the necessity of making the longitudinal seam in the packaging machine. The material should be coloured to reduce light transmission. The sequence of operations when packaging into single-service containers comprises forming the container, filling and sealing, storage of the packaged product and dispatch to wholesale and retail outlets. Here again, space must be provided for the paper or plastic stock.

Types of Dairy Products

- Milk : Liquid and Powdered
- Flavoured Milk
- Curds
- Cheese
- Ice Creams
- Cream
- Butter etc.

Dairy products packaging requires special care and attention. Different dairy products have different packaging options. But one important factor to consider is to maintain the freshness of the product. Packaging manufacturers and suppliers have come up with wide range of films, bags, laminates and equipment for packaging bulk dairy products, as well as, consumer portioned items, by considering such factors like protecting the flavors and textures throughout display and distribution.

Branding in Packaging

Packaging - integral part of brand identity. Deliver consumer expectations of

- Natural dairy taste and freshness
- Have a unique natural appeal
- Should extended shelf life.

Pouch/Sachet

- The pouch or sachets are formed from either a reeled or flat film. UV light may be used to sterilize the film. It is used to package butter, cheese, milk powder etc.
- The popular laminate for such bag is black or dark brown (to exclude UV light) or white.

- The bags may be formed from either a reeled or flat film.
- Generally it is a form-fill-seal system.
- Generally, ultra violet light is used to sterilize the films.
- The bags are heat-sealed and cut, the common
- Sequence being to bottom seal, fill, move down on sachet length, top seal and cut off.

Carton

It plays a very important role in the bulk packaging of milk. It is used for liquid, frozen and coagulated milk products. It is mainly available as preformed containers or as precut blanks or sheets ready to be formed into containers.

The cartons are the preferred medium or form for packaging milk. They are also used for *packaging* liquid, frozen and coagulated milk products. Cartons are commonly made of food grade paper coated on the inside with wax or plastics; or lined with paper, plastic films or aluminium foil; or made of laminates.

Made of **PE/paper/PE/aluminum/PE/PE**, six-layers laminated aseptic packaging material.

- Suitable for packaging milk, juice, beverage, drinks, etc
- The shelf life can reach to 8 - 12 months
- Filling volume: 125ml, 200ml, 250ml, 1,000ml
- Can be automatically filled and sealed by Tetra Brik series filling machines

The merits include maximum space utilization in vehicles, and storage; ability to carry attractive printing and convenience as a means for stacking milk at super market shelves. Retailers in the developed countries consider it the best available package for self-service selling. Cartons also play a role in the bulk packaging of milk. Cartons are commonly available either as preformed containers or as pre cut blanks ready to be formed in to containers. The carton systems in common use are as follows.

- Perga (preformed) U.K., Pure pak (precut) USA,
- Zupack, Blockpak (precut) Germany,
- Tetra pak (precut) Sweden

Barrel

Barrels are usually made of wood and are coated with wax on the inside. They are used for bulk packaging of semi-solid buttermilk, sweetened condensed milk, etc. Commonly made up of wood and coated with wax on the inner surface. Used for bulk packaging of sweetened condensed milk, semi solid butter milk / whey, butter oil etc.

Cup

Cups are made up of paper with plastic or wax coating on inner surface. They are usually covered with a lid and used for frozen and coagulated products.

Cans

This is popularly used for various types of solid, semi-solid and powdered products. Aluminum cans widely used. They are the most convenient for gas packing and suitable for dairy packaging. This is commonly used for packaging all types of solid, semi solid and powdered dairy products. Cans are traditionally made of soldered tin plate steel, generally lacquered on the inner surface to prevent corrosion. Aluminium cans have now become famous. Cans are the most convenient for gas packing.

Bottles

Made of glass with aluminum closures, the bottles are suitable for storing milk shakes and liquid stuffs. The glass bottle still continues to be used for packaging of milk in some parts of the world.

However in several developed countries and some of the developing countries it has lost ground to single service containers for packing milk.

Collapsible Tube

It is used for packaging semi-fluid products like sweetened condensed milk, processed cheese spread etc. It is usually made up of aluminum and lacquered on the inner side. They are made up of aluminium and lacquered on the inside. Low cost, lightweight, ease of handling and dispensing product protection are its advantages. Used for packing semi fluid products such as sweetened condensed milk, processed cheese spread etc.

Types of Materials Used in making Dairy Packaging

- Metalized plastic films
- PVDC coated films
- Plastic
- Paper
- Nylon
- LDPE
- Laminates
- Polyolefin
- Polypropylene
- LDPE polyethylene
- LLDPE polyethylene etc.

Milk Powder

- Tin plate containers, nitrogen packed, and lacquered from outside.
- Flexible laminates such as METPET/BOPP/Aluminium foil/Poly laminates.

- Refill packs; lined cartons laminated with BOPP/PET, varnished on the outside.
- Paperlaminated film is also used.
- Bag-in-box; Powder filled in laminate and packed in cartons.

Butter

- Duplex board with vegetable parchment paper
- Tin plate containers
- Aluminium foil

Cheese/Cheese Spread

- Tin plate containers lacquered from inside,First packed in aluminium foil and then in duplex board carton
- Injection moulded PP/HDPE container

Ghee

- Tin plate containers lacquered from inside
- Glass bottles
- HDPE film pouches

Ice Cream

- Thermoformed/Injection moulded plastic containers
- Duplex board carton (poly laminated)
- Laminates of BOPP/PET

Other Indian Dairy Products

- Injection moulded/thermoformed containers (shrikhand, gulab jamun)
- Stand up laminated pouches

Packaging of Bakery Products

The bakery industry, production of which has been increasing steadily in the country, is the largest among the processed food industries in India. The two major bakery industries, *viz.* bread and biscuit account for about 85 per cent of the total bakery products. The annual production of bakery products, which includes bread, biscuits, pastries, cakes, buns, rusk etc.,is estimated to be in excess of 3 million tonnes. The production of bread and biscuits in the country, both in the organized and unorganised sectors, is estimated to be around 0.44 million tonnes and 11 million tonnes respectively. India's bakery market at Rs. 60 billion tonnes makes it the third largest market in Asia Pacific, only after Japan and Australia. The bakery industry in India comprises of organised and unorganised sectors. The organized sector consists of large, medium and small-scale manufacturers who produce packaged biscuits and bread. The unorganised sector consists of small bakery units, cottage and household type manufacturing their goods without much packaging and distributing their goods in the surrounding areas. Bread market is estimated to be growing at around 7 per cent p.a. in volume terms, whereas the biscuit market has witnessed a higher growth at around 8- 10 per cent. Within the biscuit category, cream and speciality biscuits are growing at a faster rate of 20 per cent p.a. The per-capita consumption of biscuits in India is around 900 gm as compared to 15-20 kg. for developed countries. The consumption of biscuits is equally divided between the urban and rural population. Demand for biscuits in 2003-2004 is likely to exceed 1.2 million tonnes. Bakery products are an important source of nutrients *viz.* energy, protein, iron, calcium and several vitamins. Commercial bread and biscuits contain around 7.5 per cent to 7.8 per cent protein respectively. Biscuits are amongst the lowest cost processed food in the country when compared to other Indian sweets and salted snacks. Biscuits are easy to use during travel or at home because of its availability in variety of pack sizes. They also offer substantial energy. Thus biscuits have an important role to play as a diet supplement for both adults and children. It is no longer viewed as a luxury tea-time snack but essential daily food component for an average Indian household. The packaging of bakery products is closely interlinked with production, preservation, storage, transportation and marketing. The importance of packaging can further be gauged from the fact that packaging constitutes a fair portion (10 to 25 per cent) of the entire cost of the pack.

Product Range

Bakery products contain high nutritive value and are manufactured from wheat-flour, sugar, baking powder, condensed milk, ghee (fat), salt, jelly, dry fruits, various essences and flavouring etc. Different type of bakery products can be classified as:

Dry Bakery Products

- Biscuits: Soft biscuits, hard biscuits, cookies, crackers, fancy biscuits, cream wafer biscuits.

Moist Bakery Products

- Bread: Sweet bread, Milk bread, Masala bread, Garlic bread, Fruit bread etc.
- Buns: Fruit buns, hamburger buns, dinner rolls, crisp bread, pizza.
- Others: Cakes, pastries, doughnuts, muffins etc.

Product Characteristics

Dry Bakery Products

These products are fragile and characterised by a low moisture content (<6 per cent) low water activity (Aw = 0.30) and are highly hygroscopic. Moisture is the decisive criteria for the organoleptic properties and acceptability by the consumers.

Loss of Crispness

Biscuits have a low moisture content, high fat level and are fragile in nature. Hence, they have to be protected from these three aspects. Since the biscuits consists of wheat flour, fat and shortening, sugar, salt and flavouring agents they are predominantly sensitive to water vapour interchanges (moisture) and oxygen reactions. They generally have an initial moisture content of 2-4 per cent equilibrating to 10-15 per cent RH. The critical moisture level from the point of loss of crispness varies between 6 to 8 per cent. The shelf-life of biscuits depends upon:

- Inherent characteristics of the product
- Barrier and other functional properties of the packaging material
- Packaging operations adopted
- Distribution and storage patterns followed
- Economic considerations

There is a well-established relationship between water vapour sorptions and chemical, physical and stability characteristics of biscuits. For predicting product shelf-life and package performance with respect to water vapour transfer, the data required is:

- Water sorption isotherm
- Water Vapour Transmission Rate (WVTR) of the packaging material
- Storage Conditions

Since these are moisture sensitive products, water vapour transmission rate of the packaging material used is of importance as it is closely associated with drying, physical structure and protective action against oxidation. These products not only become brittle and hard but also develop oxidative rancidity at very low moisture contents. Temperature also plays a very important role. As the temperature increases the critical moisture gets reduced since the Equilibrium Moisture Content (EMC) corresponding to the same water activity is decreased. Products with high fat content also tend to go rancid. The recipe and processing of the product has a major contribution to the product loss of crispiness or turning rancid. The rate of loss of these characteristic depends on it. IS 1011-1992 specifies the maximum limits for moisture, acidity and acid insoluble ash for biscuits.

Rancidity

Another requirement due to high fat is the prevention of rancidity. When fat gets exposed to moisture and atmosphere, it gets oxidised and this results in rancidity and lowering of shelf-life. Fruits and nuts used are also susceptible to oxidation in presence of oxygen. Hence the packaging material must be grease resistant to prevent seepage of fat and staining of the pack and have low oxygen permeability to prevent oxidation and rancidity of the fat.

Light

Light is also detrimental to colours or cause oxidation of fats leading to rancidity producing undesirable off-flavours. In such cases, opaque packaging material is used. Some biscuits are susceptible to tainting by inks, adhesives and coatings used in the packaging material. The packaging material, therefore, should be free of residual solvents etc; to avoid development of off-flavours.

Packaging

Packaging plays a significant role in bakery products in increasing shelf life, preventing mechanical damage, marketing, nutrition value and displaying food safety related warnings. We would concentrate on types of packaging used to attract consumers or through creative packaging to enhances sales and product s uniqueness from competitors. There are several basic requirements of a package intended to contain bakery products.

These include:

- Water vapour permeability of packages
- Oxygen exchange from within and outside a package
- Aroma impermeability characteristics of packaging materials
- Resistance to seepage of fats and oils
- Protection against deteriorative visible and ultra violate radiation
- Good printability and appearance
- Physical, mechanical protection to the products against shocks, crushing and vibrations
- Compatibility and safety of the packages

Dry Bakery Products

Packaging, in general, must meet the following five basic requirements:

- ***The biscuit pack must give barrier protection to the product.***

  This can be achieved by packing the biscuits in good moisture and gas barrier packaging material. The pack must be perfectly heat-sealed in wrapping materials to protect against light, humidity and external odours. The result is a product, which is fresh and tasty throughout its shelf-life.
- The biscuit pack must give mechanical protection to the product. This can be achieved either by packing the biscuits in end-fold style portion packs or by gas flushing the pillow packs, thus preventing breakage during transport and retail handling. Ready to sell individually wrapped packs eliminates the hygiene factor risks since the biscuits do not come in contact with the external environment. The appropriate films and correct sealing prevents any infestation by insects.
- The packaging must appeal to the potential customers and stand out against other competing products and serve as an effective advertising tool.
- Detailed information about the product such as composition of the product, nutritional value, price information etc.

- Satisfy consumer demand for convenience packaging by providing different pack sizes, convenient packet opening facilities like tear tapes, incision cuts etc.

Moist Bakery Products

Breads and cakes are another category of baking products with comparatively less shelf-life. These products have high moisture content (>12 per cent), supple texture and high water activity between 0.6 to 0.85 with low resistance and tendency to crumble and go stale. Since it contains hydrated starch it is prone to staling, thus limiting its shelf-life. It has low fat content and short distribution life, hence it does not need protection against oxygen. Bread is also susceptible to loss in aroma / flavour, so the packaging material used must prevent pick up of undesirable off-flavours. The undesirable changes during storage include moisture loss, staling and loss of freshness. The packaging material must possess moderately effective moisture barrier properties. The inner portion of bread has equilibrium humidity in the range of 90 per cent, hence it tends to dry out rapidly and becomes harder. The crust however, has low equilibrium humidity and it tends to become soggy under moist conditions. Too good a moisture barrier, has effect of promoting mold growth on the bread and allows the bread to become soft. If a poor barrier film is used, the bread will tend to dry out and stale. Staling of bread starts within 3-4 days of manufacturing. This is an inherent property of the type of flour, method of baking and storage conditions. It is caused by the migration of water from the starch to the protein portion of the interior, the starch then becomes dry and looses texture. Since this activity is independent of the moisture content of the inner portion of the bread, an effective packaging material must protect the bread until staling occurs. The ideal bread packaging material must be attractive, strong and inexpensive. It must have adequate moisture barrier properties to improve the shelf-life, able to run on automatic machinery and lastly should protect the shape of the product. Hence the packaging material selected must conserve the moisture content, prevent staling and keep the bread in a fresh condition as long as possible.

The ideal bread packaging material must:

- be attractive
- maintain adequate shelf-life
- run on automatic machinery
- be strong
- be inexpensive
- be an adequate moisture barrier, and
- protect the shape of the product

Bread

Traditionally, bread in India was packed in waxed paper wrappers. The search for lower cost over wrapping materials led to the use of polyethylene film and nearly 80 per cent of all bread is now packed in plastics films such as LDPE, LLDPE-LDPE and PP. Also, autobagging machines require high slip PE resin *i.e.* pouches with good openability. LLDPE / LDPE bags of 1 to 1.5 mm thickness secured by plastic clip or twisted wire ties are normally used.

Cakes, Pastries, Doughnuts

These products are available in various sizes, shapes and forms. Since these products contain high moisture content they are prone to mould growth and hence the packaging material selected should not encourage mould growth. The packaging material used is Polypropylene (PP), Cast Polypropylene (CPP), Poly Vinyl Chloride (PVC) etc. while the choice of the film depends upon the machinability and economics required.

The Packaging Styles

There are several popular wrapping styles, which are applied widely to a variety of biscuits (of all shapes and sizes). Biscuits packed using the following two wrapping styles must be of common size and shape with a certain consistency and rather narrow tolerances in their dimensions. Standard wrapping machines can be used.

Endfold Wrapping

This wrapping style is the classic, traditional biscuit wrapper. A portion of biscuits standing on edge is roll – wrapped or fold wrapped into a heat sealable film. The longitudinal packet seal is sealed tightly in a fin seal style. The packet ends are folded neatly and heat-sealed. Due to the neat and tight surrounding of the film, this packet gives utmost mechanical protection and acceptable barrier properties for hard and semi -hard biscuits and many other cracker types. Enfold wrapping is considered the most effective in terms of presentation by many marketing specialists - not only due to neat and impeccable shape, but also due to its ability to clearly distinguish the product amongst the host of pillow pack items on the retail shelves.

Pillow Pack Wrapping

This is the standard wrapping style for smaller biscuit packs (snack packs/ single serve packs) containing one or more piles of biscuits. In addition, pillow pack wrapping is used for bigger packets with products standing on edge (Slug wrapping) as well. In this configuration, it often serves as a primary wrapper, to be over-wrapped by a carton toimprove presentation and acceptance. The main advantage of pillow packs on edge, is its flexibility with regard to the slug length. For instance, it allows the machine to automatically adjust the length during wrapping by means of tendency controlled check weighers. This feature ensures thehighest weight accuracy. Additionally, the pillow packs typical fin seal style sealing is somewhat tighter than the enfold wrap. This disadvantage of pillow pack slug wrapping is its limited mechanical product protection due to its rather loose packing. Further, the presentation of products packed using the pillow pack style is considered by most to be less attractive than enfold packets. Packing for Odd-sized Biscuits

Summary

Bakery products include items of different packaging requirements, which are met by a range of plastic materials in the form of films, laminates and thermoformed trays. These materials provide adequate protection against moisture loss / gain, retain the taste and aroma, and are hygeinic and safe for food contact. Other additional properties such as machinability, printability and cost effectiveness make them the ideal choice for a package.

Packaging of Confectionery and Chocolates

Confectionery is the general term applied to various varieties of sweets and chocolates. Confectionery has four basic ingredients: Sucrose, Glucose, Invert Sugar and water. Alone these confections are called boiled sweets or fondants. By adding further ingredients like milk solids, fat, protein and modified starches, range of confections include jellies, pastilles, toffees and caramel. By nature chocolate is a different product from the other confectionery. Chocolate is simply described as a suspension of finely ground sucrose in cocoa butter and a wide variety of confections can be obtained by combining all these various products. Chocolate is a food product prepared from ground and roasted cocoa beans(rich in fat, carbohydrate and protein). It is a valuable source of energy for those who need to be physically active like soldiers and athletes. It also contains small amount of caffeine, mild stimulant. It is made by roasting cocoa beans, removing shells. Kernels ground to powder, which turns into liquid rich in fat due to heat of grinding. This is called chocolate liquor or cocoa butter. This is the base for all chocolates. Various types are made by adding sugar, milk powder, vanilla and other ingredients. It is very susceptible to temperature, the melting point of this fat is 36°C.

If melting subsequent solidification takes place the surface is rough textured. A grayish discoloration takes place if temperature oscillates below melting point, caused by fat coming on to surface.

The chocolate types is divided into three main segments:

1. Solid Chocolate slabs, bars and novelties
2. Filled bars with soft and hard cereal centres.
3. Chocolate assortments.

Boiled sweets and Toffees are made by boiling constituents into molted viscous mass which is formed into a continuous hot rope, chopped into pieces, punched into shapes and immediately wrapped. Unwrapped pieces could stick to each other and deshape. Peppermints are a little more than flavoured finely ground compact sugar. These are quite stable at ambient conditions and need little protection by way of packaging. The moisture content of confectionery will range from 0 per cent in chocolate to 23 per cent in jellies. Organised market for sugar confectionery is estimated to be 1,39,000 tonnes per annum and is growing at the rate of 10 - 15 per cent per annum. The confectionery market has undergone a metamorphosis in the last few years. From the commodity market controlled by local players, it has changed to a branded products market with strong presence of multinational companies. The confectionery market is highly fragmented with several players with strong regional presence. The entire sugar confectionery market can be divided into eight major categories, *viz.* chocolates, hard-boiled candies (HBCs), toffees, eclairs, chewing gum, bubble gum, mints and lozenges. Chocolate market can be segmented into moulded chocolates, count chocolates, panned chocolates, eclairs and assorted chocolates. Moulded chocolates are the largest segment accounting for more than one-third of the market.

Product Characteristics and Packaging Requirements

The key raw materials for sugar confectionery are sugar (60 - 65 per cent), glucose, citric acid and flavoured essences. Confectionery is hygroscopic in nature and requires protection against the ingress of moisture, and exposure to high temperature (as far as possible). Temperature during storage can be critical, but packaging alone cannot normally control the heat reaching the product. Obvious protective functions like prevention of moulds, yeasts, foreign odours and mechanical damage and in some cases light must be borne in mind. Chocolate coated items need little in the way of moisture barrier and chocolate coated wafer, for example will remain crisp and fresh in a non moisture proof wrapping, provided temperature during storage does not permit the chocolate to melt or soften. A candy or confectionery product may be adversely affected by many things. Hard candy, brittles and crunch products are most sensitive to moisture and absorb water vapour fairly rapidly from the atmosphere.

Because of the sensitiveness of chocolate to temperature in tropical countries, aluminium foil and metalized film substrates are popular to reflect heat and provide fresh appearance. Generally marketing requirements demand high quality presentation for these products which are not cheap. Gravure printing of designs is almost universally used for chocolate on substrates which show graphics in best possible way. For this reason, foil or the foil appearance materials are used.

Foil has another important function – if the chocolate should become molten and later solidify, the smooth inner surface and release properties together with the rigid deadfold characteristics disguise the fact that the chocolate ever melted at all. For this reason also, chocolate generally is not wrapped in transparent materials, such as cellophane unless in the case of chocolate assortments, where individual chocolate coated sweets would be individually film or foil wrapped.

The traditional wrapping for chocolate bar is an aluminium foil inner layer(0.012mm foil – dull side inwards – bright side outwards) a glassine layer over the foil to protect it and printed coated bleached sleeve of approximately 90 GSM. The wrapping machinery to do this pack would take all dead fold characteristics to keep the layers closed. Cold seal coatings are given for the sealing the layer in contact with chocolate. The cold sealing areas are applied on the reverse side to register with the printing on the front such that they only come in contact with each other when the wrapping is applied at the time of printing using gravure or flexo techniques. The main advantages of cold sealings are;

- No possibility of melting chocolates when the machine is stopped or jammed.
- Machine speed is variable between wide limits
- No start up delay waiting for heat seal jaws to heat up to operate temperature. These advantages to user offset its higher price.

Chocolate assortments are invariably the various chocolate coated fillings which are foil wrapped in trays in quality printed cartons with a cellophane overwrap to enhance presentation and keep foreign odours away. Important functions of a chocolate slab wrappings are:

- Attractive graphics
- Hermetic closure – for heat and foreign odour protection
- Rigidity for good appearance

The printing on the outer substrate is generally sandwiched printing between transparent outer layer and inner sealing layer.

The sugar component of most of the types of confection is in a soluble form and it is essential that this form is maintained, since crystallization of sugar alters taste and texture giving the impression of an old product.

The moisture content of confections must be retained at the level during manufacture and this is the chief function of wrappings. Moisture can be given or taken up from atmosphere. The Table 1 of ERH shows the details. Items with ERH less than 30 per cent pick up moisture and so need a high barrier. As the ERH increases it is progressively less critical. Other factors of spoilage are heat, light, moulds, yeast, foreign odour and mechanical damage. Boiled sweets and Toffees are made by boiling constituents into molted viscous mass which is formed into a continuous hot rope, chopped into pieces, punched into shapes and immediately wrapped. Unwrapped pieces could stick to each other and deshape. The important function is to keep pieces separate. In order to keep moisture a moisture proof materials is normally applied as a twist wrap, fold wrap or bunch wrap. A cellophane or BOPP overwrap is with a waxed paper centre strip is popular for toffees to retain the square cross section of the confection.

Twist wrapped confections would be weighed into a bag either pre made or on FFS machine. The main function of the outer bag is for moisture protection and better presentation. Twist wrapped confections in VFFS outer bag provide good cushioning effect which will keep the sweets in a presentable condition. Various gums are less sensitive while some creams tend to lose moisture considerably. The factors that lead to spoilage of confectionery are highlighted as below. While the range is very wide,. the main packaging requirement is protection against either moisture ingress or egress. In other words the packaging should be moisture proof.

Crystallisation

The sugar component of most types of confection is in a soluble form and it is essential that this form be maintained, since crystallisation of the sugar alters the taste and texture giving impression of an old product. Moisture interchanges play a major decisive role in deciding the shelf-life of confectionery items. The Equilibrium Relative Humidity (ERH) of a confection during its life determines its sensitivity towards all physico-chemical changes that occur to environmental conditions. Data on moisture sorption characteristics of a confectionery item are very important in understanding the storage stability as regards its chemical and physical changes, growth of microorganisms and also drying characteristics and product formulation and package selection.

The ERH of confectionery products are given in following Table

Equilibrium Relative Humidities of Confectionery

Confection Type	Type of Deterioration	ERH (%)
Boiled Sweets	Graining and Stickiness	< 30
Toffees(caramel type)		< 50
Gums and pastilles	Stickiness, micro-organism growth	65
Marsh mallows		65 - 75
Marzipan	Drying out and mould growth	70 - 85
Jam		75 - 85
Milk chocolate Syrup	formation and sugar bloom	80
Plain chocolate		85

[Source: Minifie 1988]

Oxidation

If the confectionery contains fat, then rancidity can occur as a result of oxidation. Fat containing confectionery will also react with moisture to produce fatty acids and their degradation products. Oxidation of fats in chocolates is a minor problem as coca butter contains natural antioxidants. Being high in fat, however, chocolate is likely to absorb odours from the surrounding atmosphere.

Fat Bloom

It is a surface defect in chocolate, which occurs during storage, whereby the initial gloss of the chocolate is first lost and then replaced by a white or grayish haze. This defect, occurring more frequently in summer, is deleterious to the aesthetic appearance of many chocolate products. The defect, however, does no harm to the eating quality of the chocolate, unless it brings along with it other defects such as staleness or mould growth.

Sugar Bloom

Chocolates can be affected by condensation giving sugar bloom, in which a fine layer of sugar crystals form on the surface of the product. This renders it unsaleable and if left unchecked can lead to mould growth.

Odour Absorption

Odour absorption by oils and fats can also be a problem. Odourous compounds are often very soluble in oils and fats and can be readily absorbed from materials such as paints, printing inks, petroleum oils and disinfectants. When the product is eaten, the odours are released in the mouth producing objectionable flavours. Confectionery can also deteriorate in other ways including exposure to heat, light, moulds, yeast, foreign odours and mechanical damage.

Selection of Packaging Material

In selection of packaging materials for confectionery the following need to be considered:

Water Vapour Transmission Rate (WVTR)

Knowledge of WVTR of packaging materials and the effect of folding, creasing, crumpling of materials on papers and aluminium foil show considerable effect. However, thermoplastic materials are not much affected. Following Table gives the effect of folding, crumpling on WVTRs of some thermoplastics.

Effect of Folding and Crumpling on the WVTR

	WVTR, g/m2, 24 hr. 38°C & 90% RH.			
	Flat	Folded	Crumpled	Gelboflex
Met PET (12μ)	0.9	1.7	3.4	18
Met PET / LDPE (50μ)	0.5	0.6	0.7	0.6
2-sides PVDC coated PET	4.2	3.7	5.9	4.8
2-sides PVDC coated PET/LDPE	2.9	3.6	3.3	3.1
[Source: Dulin (1978)]				

Gas Transmission Rate

Permeability to gases like oxygen decides the shelf-life of oxygen sensitive confectionery items.

Plastic Gift Packs for Chocolates

Besides OTR, permeability to volatiles and flavours is important in confectionery packaging. Polyolefins have high values, whereas plastics such as polyester, nylons, ethylene vinyl alcohol (EVOH) have good barrier properties for transmission of volatiles.

Grease Resistance

Grease Resistance of the packaging material is important to avoid seepage of oils and fats and smudging of the print.

Tensile Strength and Elongation

Tensile Strength and Elongation properties of materials need to be studied as their running on high-speed machines should be suitable.

Tear Strength

For a confectionery film, tear strength is of importance as low tear values are necessary and useful for opening packages by hands.

Heat Seal Strength

The performance of a finished package is determined by the effectiveness of the package seal *i.e.* the permeabilities to water vapour, gases and volatiles increase if the seal is not perfect. Thermoplastic films such as polyethylene give excellent heat seals.

Performance Properties

Apart from the above mentioned important properties, a material has to perform well on machines, therefore knowledge of physical properties like slip, stiffness, blocking resistance is also necessary. Twist retention for twist wrap is also of importance. The initial function of packaging is to protect. However, the emotional role played by packaging is also of importance, especially when the confection is a gift. A sophisticated packaging using deluxe materials is often used as a way of expressing feelings. Confectionery packaging must also be specialised for specific target groups. Children's sweets are to be packed differently from adult sweets and chocolate bars for adolescents should look different from expensive chocolates for discerning consumers. A different pack size is required for quick impulse buys at petrol stations and roadside shops than for the super markets selling predominantly family sized packs.

Packaging Materials used for Confectionery Items

A very high quantum of polymeric materials, besides cellulosics and aluminium foil are used for confectionery items. Paper board and metal containers are also used for certain applications. Although a variety of packaging materials are available, the ultimate choice of the wrapper depends upon the required shelf-life, performance on the wrapping machine and the cost which is purely based on the segment of the market targeted by the manufacturer. The most common choice of packaging medium is plastic (generally flexible) as it provides the required protection and preservation, grease resistance, physical strength, machinability and printability. Plastics being lighter in weight are, therefore, the most preferred material for packaging of confectionery. There are many changing trends in the packaging of confectionery. Plastic films and their laminates are increasingly replacing waxed papers due to better properties and aluminium foil laminates due to price and better flex crack property. Depending on the type of package *i.e.* twist wrap, pillow pack and vertical flow pack or roll pack, the plastic based packaging films used for confectionery are listed below.

Polyethylene (PE)

It is considered to be the backbone of packaging films. Since one of the greatest threats to the integrity of confectionery products comes from moisture, polyethylene with its low water. Chocolates Packed in Heat Sealed Laminated Plastic Films, Foil Wrapped Chocolates with Paper Board Cartons vapour transmission is of definite interest. Polyethylene films are fairly free of plasticizers and other additives and are quite extensively used as a part of lamination. Its ability to heat seal increases its value.

Low Density Polyethylene (LDPE) is an economical material with low WVTR, however, it has high permeabilities to flavours / volatiles, poor grease resistance and are limp. High-density polyethylene (HDPE) is stiffer, more translucent and has better barrier properties but needs higher temperature for sealing. Later additions include high molecular weight high-density polyethylene (HM HDPE) and linear low-density polyethylene (LLDPE). HM HDPE is a paper like film with high physical strength and barrier properties, but is less transparent than ordinary polyethylenes. HM HDPE is available in twist-wrap grades. Polyethylene films are also suitable for making bags and pouches. A copolymer of polyethylene and poly vinyl alcohol, and EVOH has outstanding gas barrier properties specially when dry.

Polypropylene (PP)

Polypropylene films are undergoing a growth trend in the confectionery industry. They have better clarity than polyethylenes and enjoy superior machineability due to stiffness. Lack of good sealability has been a problem; however, PVDC and vinyl coating have been used to overcome this problem. Some varieties of PP have been specially developed for twist-wrap applications as they have the ability to lock in position after twisting. Pearlised polypropylene with an opal finish and attractive gloss is also used. Both as laminates and overwraps, PP film is now widely used for all types of confectionery packaging applications.

Poly Vinyl Chloride (PVC)

PVC is a stiff and clear film having low gas transmission rate. PVC can be used as smalln wraps, bags and pouches. PVC when co-polymerised with polyvinylidene chloride is known as Saran. Since it is a costly material, it is only used as a coating to obtain barrier properties and heat sealability. PVC film is also used for twist wraps, as it has twist retention properties and is excellent on high-speed machines.

Polyesters (PET) and Polyamide (PA)

Polyethylene terephthalate film has high tensile strength, gloss and stiffness as well as puncture resistance. It has moderate WVTR, but is a good barrier to volatiles and gases. To provide heat seal property, PET is normally laminated to other substrates. Nylons or polyamides are similar to PET, but have high WVTR.

Metallised Films

When polymeric films are metallised there is an improvement in their barrier properties. Metallisation is also used for decorative purposes and aesthetics. The films, which are used for metallisation are PVC, PET, PP and polyamides. Hard-boiled sweets are generally twist wrappedindividually in twist retaining plastics such as Poly Vinyl Chloride, Polypropylene and High Density Polyethylene. To provide greater protection against water vapour ingress, a secondary pouch made of LDPE or PP containing 100 grams to 1 kg of the product may be used. Hard-boiled candies are also packed in tuck-fold heat sealable films. Medium hard sweets and chewy candies such as caramels and toffees are either twist wrapped or packed in heat sealable tuck-fold wraps. Special grades of HMHDPE, PP and PVC are used for this purpose. For gums, moderate moisture barrier is required, "sweating" is to be avoided and a slight liberation of moisture is necessary. The possibility ofcondensation is avoided by polyethylene. The traditional wrapping for the chocolate slab is aluminium foil for the inner layer (0.012mm) with the dull side inside, a glassine layer over thc foil to protect it and a printed coated bleached kraft sleeve. A modification of this style of chocolate bar wrap is to use a heat seal coated foil inside the paper sleeve. Securing of the paper sleeve is by hot melt adhesive. Another popular method of wrapping chocolate slabs is on the horizontal form-fill-seal (HFFS) machines, when the reel of laminate is wrapped around the slab, cut and heat sealed at both the ends and along a fin-style back seam. Typical constructions are:

- 12μ Polyester / 9μ Al foil / heat seal lacquer
- 40 gsm coated bleached kraft / 9μ Al foil / LDPE plus cold seal coating

Chocolate bar lines are mostly individually wrapped on the horizontal form-fill-seal (HFFS) machine. The typical wrapping materials are Polyester, OPP - pearlised or metallised. The Indian Standard Stipulates that confectionery shall be wrapped in plain or printed cellulose film, waxed paper, aluminum foil, polyethylene or other flexible packaging materials. In the case of printed packaging materials, it is specified that the printing ink shall not come in direct contact with the product. To safeguard the interest of the consumer, the Standards of Weights and Measures (Packaged Commodities) Rules, have imposed a limit on the weight of the wrapper. Under this, it is essential that in the case of twist wrap and pillow wrap confectionery, where the weight of the individual pieces is less than 10 grams, the size of the wrapper and the type of wrapping material selected is such that it meets the limit on weight of wrapper, under the above rules.

Future Trends

The confectionery market is one of the most competitive in the FMCG area. Major companies continuously battle to entice sweet-toothed consumers from competing brands. A strong brand identity is complimented by innovative packaging designs to deliver a product capable of meeting the consumer demands.

The developing trends in confectionery packaging are:

- Widespread and increasing use of cold seal
- Use of laminated structures and cold seals for premium products
- Increasing use of opaque multi-packs for grocery outlets
- Switch over to higher yield opaque films for cost reduction
- Replacement of Al foil / paper wraps by OPP laminates
- Developments in low temperature heat seal packs

Emerging trends in the confectionery and chocolate packaging industry have given wide scope for development of a variety of innovative packaging media depending on the required shelflife and performance of wrapping machines. Plastic films and laminates are the most popular choice as a packaging media, replacing traditional waxed paper and aluminium foil. Polyethylene is considered as the backbone of packaging films along with other polymers like PP, PVC, PET and PA. The ability of plastics to pass all selection criteria as an effective packaging media has led to very high quantum of polymeric material used for confectionery items

16

Packaging Material Compatibility

What is Compatibility?

"COMPATIBILITY" is a word, which is very commonly used in our day today life. One can sight a lot of examples of compatibility and non-compatibility if one tries to think the happenings in one's life. Some of the examples are:

1. Divorce of a wedded couple-because of non-compatibility
2. Not cooking in copper vessels without tinning-because of non-compatibility
3. Pickles are not stored in metal containers – because of non-compatibility
   The list is never ending. Then what is compatibility?

It is the ability to co-exist (without affecting adversely each other). A compatibility study is the studies related to the understanding of the mechanism of co-existence. Compatibility in the context of packaging technology is the studies on compatibility of product and packaging keeping the environment in view.

In order to understand the importance of compatibility studies it is better to study some cases.

CASE 1 – SCALPING

CASE 2 – OVERALL VISION

CASE 3 – BIG PROBLEM, SMALL PACKAGE

CASE 4 – SHEDDING SOME LIGHT

CASE 5 – OFF-FLAVOUR CASE 6 – ODOUR

CASE 7 – TURBIDITY

CASE 1- SCALPING

PRODUCT: ORANGE JUICE

PACKAGE: "Juice Board" *i.e.,* LDPE/Paper Board/LDPE COMPLAINT: Weak Flavour and Turpentine like Off-Flavour

CHECKED: Variation from Different sources and Processing Temperature

Complaint samples contained low levels of flavour important chemicals, but high levels of terpeniol with lemonine oxidation. Analysis of the packaging material revealed most of flavour imparting chemicals absorbed into the by the board material. This is called *SCALPING.* Further analysis of orange juice packaged in.

! Juice Board (LDPE/BOARD/LDPE)

! Barrier Board (LDPE/Board/LDPE/EVA Co polymer

Indicated that absorption and oxygen transmission in Juice Board caused the problem. In Barrier Board the juice did not have off flavour. Hence high barrier material was needed to resolve the problem Unfortunately, the juice division, which had experienced long term sales dump, was SOLD OFF by the time the solution was worked out.

CASE 2 - OVERALL VISION

PRODUCT : HIGH FAT SANDWICH SPREAD PACKAGE : Tubs and Lid with PET lined wad COMPLAINT : Product rapidly TURNS RANCID

CHECKED : Barrier properties of PET lined wad and also the more expensive and better barrier FOIL lined wad. Organoleptic tests of both the wads was also done Tried Lid with FOIL lined wad (Expensive) Product turned RANCID in packs with both the type of wads

CASE 3 – BIG PROBLEM, SMALL PACKAGE

BACKROUND

A Food Products Company wanted to do a sample promotion a fried cereal product and decided to do with small packages. They used EXACTLY SAME barrier film packaging material as they successfully use for their larger packs for many years. Shelf Life testing seemed not necessary. WRONG !!!!!

The small packs quickly developed oxidation off-flavours and the promotion was an embarrassment and big waste of money

WHAT WENT WRONG?

Here the team working on this project knew which barrier to use. BUT did not know the proper BARRIER TO PRODUCT RATIO to use. A stronger BARRIER film was required because of the larger area-tovolume ratio with the smaller packages.

CASE 4 – SHEDDING SOME LIGHT

PRODUCT: COOKIES and ICE CREAM

PACKAGE: Clear Transparent Round Container

COMPLAINT: Putrid Off-Flavour of products. Problem isolated to one regional warehouse reoccurring UNPREDICTABLY for several weeks. WHAT WENT WRONG? OBSERVATION: the problem cropped up only for a small per cent of a day's production. Observed only in samples from top and sides.

SUSPECTED CAUSE: Storing near incompatible, odiferous goods in the vicinity and absorption volatile organics.

RECHECK: Goods stored separately. Further analyses indicate oxidation and that too light induced. With the new clue rechecked storage. Pallets of ice cream stored near the HIGH INTENSITY warehouse lights.

CASE 5 - OFF FLAVOUR

PRODUCT: QUACKER OATS (Ready to Eat Cereal)

PACKAGE: Paperboard Carton with glassine liner

COMPLAINT: Off-Flavour - Resinous odour WHAT WENT WRONG?

ACTION: Analysis for OFF FLAVOUR in the Ingredients, Process, Packaging and

Environment was carried out, by Organoleptic methods and instruments. *NO OFF FLAVOUR detected in ingredients and process*

Packaging Material foe the outer box did not have off flavour. But inner layer of GLASSINE was giving this offodour. The material was board/resin/glassine structure. The resin used for the bonding was giving off the odour

CULPRIT : RESIN used in bonding

ACTION: Changed the bonding resin to a new one without odour.

CASE 6 - ODOUR

PRODUCT: DIARY CREAMER

PACKAGE: Foil laminated pouch (3g packs)

COMPLAINT: Musty odour like in a cat litter box. Noticed while opening the shipper. WHAT WENT WRONG?

ACTION: Checked the product for any off-odour. NONE WHATSOEVER.

Further checking of all the packaging materials indicated that printing on the pouches was giving this odour. Enquiry revealed that new type of ink was used for the NEW Graphics on the pouch. It contained TRIMETHYL BENZENE, which gave the musty odour. Foil prevented odour penetration and product was safe

CULPRIT: INK used in printing

ACTION: Changed the ink to a new one without TMB.

CASE 7 - TURBIDITY

PRODUCT: CLEAR TRANSPARENT LUBE OIL

PACKAGE: Lined Carton. Paperboard outer. Liner of Co ex LDPE/HDPE COMPLAINT: Clear solution becomes TURBID

WHAT WENT WRONG?

STUDY: The product becomes TURBID with moisture ingress. Hence check was made for moisture ingress. NONE WHATSOEVER

Further product kept in Glass Bottle (Standard Sample of Batch) was clear. Chances of moisture transmission thru liner was suspected and checked. NO CLUE. In GLC of the Standard and Complaint sample revealed certain peaks in the complaint sample. Analysis of the liner also showed similar peaks. Process Chemicals and Additives used in extrusion of the liner were checked and analyzed. They also showed similar peaks.

CULPRIT: Additives used in extrusion

ACTION: Avoided the use of such additives in extrusion.

These seven case studies indicate that the cause for incompatibility can lie anywhere. Only a thorough systematic analysis can help in identifying the real cause and eliminating this problem.

Summary

Hence compatibility studies involve:

- COMPREHENSIVE PRODUCT TESTING
- COMPREHENSIVE PACKAGING TESTING

Keeping in view the overall activities of processing, Storing, Distribution.

PRODUCT TESTING Involves understanding of:

- Product
- Product Ingredients
- Product Processing
- Product Characteristics
- Product Sensitivity
- Critical Variables
- Product Testing

PACKAGING MATERIAL TESTING Involves understanding of:

- Material of Construction
- Additives and Coatings
- Material Processing
- Inks, Solvents and Adhesives
- Material Characteristics
- Material Sensitivity
- Material Critical Variables
- Material Testing

PRODUCT TESTING TECHNIQUES

- Monadic Test
- Sequential Monadic Test
- Paired Comparison Test
- Repeated Paired Comparison Test
- Triangular Test
- Ingredient Screening
- Product Optimization

INGREDIENT SCREENING

- Preliminary step in optimizing food formulation
- Understand importance of ingredients
- Helps in creating high and low concentration of certain ingredients keeping others constant
- Product evaluated overall and specifically (sweetness, mouth-feel, texture etc.)

PRODUCT OPTIMIZATION

- Improve product for max consumer satisfaction
- Structured process

PKG. MATERIAL TESTING TECHNIQUES

- Mechanical Properties
- Odour Pick Up
- Appearance
- Texture
- Aroma Retention
- UV Light
- Inks and Adhesives
- Coatings
- MVTR
- OTR
- Scalping
- Additives
- Colourants
- ESCR

17

Package Labels and Lebelling

As per Oxford dictionary *"Labels are a slip of paper, cardboard, metal etc. for attaching to an object and bearing it's name, description, destination etc."*

As per Packaging concern, Labels are usually a slip of material can be paper, filmic, metal or laminate, carrying either printed or coded information, which can be affixed either permanently temporarily to almost any type of product or packaging or service.

Labels as an integral part of the package, has to serve certain key functions:

- Attract attention
- Project the brand image
- Not be offensive to the user
- Provide useful information to the user
- Provide statutory information
- Stay with the package in all conditions
- Provide tamper – Proofness
- Give security / authenticity to the product.

It is obvious that the choice of the label material and its graphic design need to take into account all these factors. Paper, coated paper, Metallised paper, a variety of plain and opaque plastic films, aluminium foil, and fabrics are some of the commonly used materials for label applications. Since their invention, labels have undergone a sea change – from the basic wet glue type to the sophisticated self-adhesive label, and from paper to advanced film based labels. This major shift follows the recent developments in packaging, which has been totally transformed with the entry of plastics. The most prominent shift is from paper labels to plastic labels, which offer improved gloss, graphics, opacity, water resistance, and abrasion and wear resistance, durability, and the 'no-label' look.

Wet Glue Labels

Glue or adhesive is applied on the pre-printed labels just before they are transferred to the surface of the container. Lithography used to be the most popular method of printing these labels and thus the name lithographic labels. These are also referred to as 'cut and stack' labels as they are cut to size after printing and piled in stacks.

A lithographic (offset printed) label is typically the lowest cost method for printing a label and is often printed on plain paper or metallized paper. The labels are 'finished' using automated off-line methods for die-cutting, embossing, hot-foil stamping and other speciality processes. Once printed, the label is applied to the container by cold or hot glue adhesive. The adhesive can be applied either to the label or to the container. A variation is the pregummed label which when moisten adheres to the package surface.

Self-Adhesive (Pressure Sensitive) Labels

A self- adhesive label is a sandwich of facestock (paper / plastic film), adhesive, and release coating (silicone) backing or liner. The first self-adhesive product was an innovation of Stan Avery, the founder of Avery Dennision, in 1935. Today, the self – adhesive label is recognised as a high technology world class packaging medium used across a broad range of industries. Self-adhesive labels are among the most versatile labels in the world and are ready for immediate application. The labels are pre-die-cut and are individually attached to a silicone treated liner, which protects the adhesive. Owing to these advantages, self-adhesive labels have become popular across numerous segments. Using virtually every printing process available, these 'ready to apply' labels can be affixed to any substrate (glass, plastic, metal, and paperboard) because of the infinite variety of adhesive structures. Self-adhesive labels also find special applications like cold temperature and freezer applications, food contact adhesives, adhesives which can withstand sterilization and pasteurization processes and price marking.

Label applicators for pressure sensitive labels cost less than wet-glue paper label applicators. As the demand for shorter runs continues (caused by brand proliferation, onpack promotional requirements, and 'Just in Time' delivery), pressure sensitive labeling is well positioned to be the lowest total applied cost alternative. However, owing to their complicated structure, self-adhesive labels are more expensive compared to glue-labels. Self-adhesive labels have made rapid inroads in the Indian market and some of the principal applications have been in the personal care, pharmaceutical, edible oil, lubricants, beverages, food, consumer and industrial electronic products and automotive sectors.

Almost all types of printing processes are used to print on pressure sensitive labels-of these, the most popular ones are flexography and rotary letterpress.

Currently, variable imaging on pressure sensitive labels is available in the market, which permits printing unique information on each label. The alphanumeric characters can either be in a series or in random order. For example, a brand manager can now place a unique pin number or every label using digital technology. Clear pressure sensitive labels provide the package with the 'no-label' look allowing the customer to see the colour and unique characteristics of the product. The clear label can be applied to both sides of the package creating a brilliant, eye-catching, three-dimensional look. In recent years, clear pressure sensitive labels are becoming the most popular way to decorate a package. It is also a very durable label that can withstand harsh environment.

Heat Shrink Sleeves

Shrink sleeves are usually made from a monoaxially stretched PVC or PP film; the film is reverse printed, cut and then welded to make the sleeve. These sleeves are slipped over the package, manually or by machine, and passed through a small tunnel in which heat is applied. The sleeve then shrinks over the package and the specially stretched film ensure that the printing does not get distorted. The switching cost associated in the changeover of these labels is comparatively low.

This technology enables 360 degrees coverage and takes care of the varying contours of the container. Shrink sleeves can also be used as a pilfer proof device simply by stretching the label across the length of the container. They can be used on glass, PET, plastic composite cans and tin cans.

Heat Transfer Labels

Heat transfer labeling is a decorating process by which a printed design is transferred, from a treated paper carrier sheet to a plastic container under controlled heat and pressure, using controlled heat and pressure, using special application equipment. The labels are printed on a wax-coated paper in upto 6 colours, using rotogravure printing. Heat transfer labels are commonly used where flexing is important, such as with squeeze bottles. Film-based substrates are normally used for heat transfer, the content of the label being printed in reverse. The image is transferred to the container surface using a heated platen, which presses the printed ink area f the film label against the heated container. These labels can be used on commercially used plastics like LDPE, HDPE, PP, PVS, PS and also bottles of various sizes and shapes.

Metallised Paper Labels

A Metallised paper label is a metal coated high wet strength paper, printed with vibrant colours which when stuck on the container with the right glue, and gives a durable and attractive label. High labeling speeds (up to 700 – 800 BPM) can be used, with labels of different sizes and shapes. They are suitable for labeling cold beverages, beer, confectionery and food products. Metallised labels are used as an effective sales tool, especially for product where children influence the buying decision.

Security Labeling

Security labeling helps to get rid of counterfeiting and product fraud. Brand protection devices like holograms and kinetic grating are types of diffractive optically variable devices (OVDs). Batch tracing of consumer goods using a holographic label is possible using holographic barcodes, created from white diffusion patterns that emulate a printed code. The latest developments include demetalisation, which offers intricate design shapes and fine line features in transparent and semi transparent and semi transparent overlays. Adding UV florescence to the adhesive offers a second line of defense.

Tamper-Evident Lables and Seals

These have become mandatory in most retail products. These labels have security edge cuts that cause the label to tear if removed. Holographic films coated on brittle material give more protection and may have customized designs combined with non-transferable materials. Higher end devices can include hidden messages, which users can activate with pens containing special chemicals or simply by rubbing.

Electronic Article Surveillance (EAS)

This involves the incorporation of electromagnetic, radio frequency, microchips, or other data inlays into labels, tickets or tags to help avoiding retail theft. Radio Frequency Identification (RFID) systems, another technology development, which can exist with bar-coded labels, are difficult to copy or deceive and help protect good against theft and counterfeiting.

Holograms

Holograms are multi-coloured, three dimensional image-making that can be easily recognised by the naked eye. Holography is the expertise of using laser technology to generate 3D images with changing colours on a two dimensional surface. Silver or any other coloured foil is used to make these holograms. The distinct features of holograms are that they are not viable to scan, photocopy, re-print and tampering (once pasted cannot be reused). They display all seven colours of light that change when viewed from different angles, holograms increase the brand value of the product. It is mandatory for all authentic hologram manufacturers to be registered with the Hologram Manufacturers Association of India (HoMAI) and International Hologram Manufacturers Association – UK (IHMA).

The different types of holograms include :

- 2D/3D single channel which have 2 or 3 layers and no switching effect.
- 2D/3D multi channel having 2 or 3 layers and switching effects, which make visible different artworks when viewed from different angles.
- 2D/3D Kinetic where the design or pattern will be animated when the viewing angle changes.
- Dotmatrix where – 'kinetic' movements are given a dotmatrix effect. They are brighter when compared to the conventional holograms.

- True 3D hologram, which is a scaled down model of 2D artwork. When line arts are incorporated to this hologram, it gives 2D3D holograms.

For high levels of security, other effects that can be incorporated in holograms are:

- Hidden information, which can be viewed only with the help of a laser pointer
- Micro texting, which is visible only through a high power magnifying glass
- True Colour, where the colours change with the change in viewing angles.

Barcode Labels

The barcode is an automatic identification technology and is the simplest, most accurate, and most economic method of data entry. The global bar-coding business is estimated at around USD 3300 million (2001-02) and is expected to grow at 20 per cent per annum. There are various types of barcode encoding which are specific to the industry and the product. A barcode consists of a leading and trailing quiet zone, a start pattern, one or two check pattern and a stop pattern. The main purpose of a barcode is to identify the product and associated details like price, etc. Some of commonly used substrates for barcodes are thermal transfer papers and films. Major advantages of bar-coding are unique product identification, accurate stock management, and speedy and accurate consignment tracking and clearance.

18
Package Printing

Introduction

In this Chapter we are going to see what printing is, along with the terms involved with the printing press and the configurations of the printing press. We must also know the principles of every printing process which is also discussed in the chapter. Note that we shall touch upon these principles only because they help us to understand the complex machinery and the different processes actually involved in printing. In other words master the basic in order to understand the processes and the machines well.

What is Printing?

Printing is a technique whereby identical images are reproduced in quantity through a medium of printing ink and with the use of printing plate made from letters, photographs, drawings, or a combination. The duplicated image is a fixed perception record for the transmission of information. Hence, the printing image needs to have beauty, durability, safety and economy, as well as exactness.

The basic elements of printing are:

- The original
- The Plate
- The Printing Ink
- A Printing Medium, *i.e.* Substrate
- A Printing Machine

Let us now see a short description of the same,

1) The Original: is the actual data that is to be reproduced, *i.e.* printed and it can be of the form of Photographs, Drawings or Text or a combination of all the three. Since, the Original is to be printed; the data on it has to be clear and legible enough so that it can be properly reproduced.
2) The Plate: is the intermediate member for the process of printing. The data on the original is digitally, chemically or mechanically (or a combination of the three is used) transferred on to the Plate. The plate material and plate making method varies from process to process. The common materials include Steel, Rubber, Photopolymer, Tin sheets, etc.
3) The Printing Ink: is another main component which is mainly used in the printing process. The ink is used to transfer the data on the plate onto the substrate. This again has many constraints and few of them include: - Printing Process, Substrate, Viscosity, Drying Method, etc.
4) The Substrate: is the material on which the printing is to be done. The most common substrate used is paper. The others include Plastics, Corrugated Fiberboards, Textile, etc.
5) The Printing Machine: The machine is the main component for printing. Without the machine all above discussed will be useless. The printing machines are based on certain principles and these principles are discussed later. The common well known machines are made by Heidelberg, Germany. Other machines also come from Mitsubishi, Japan.

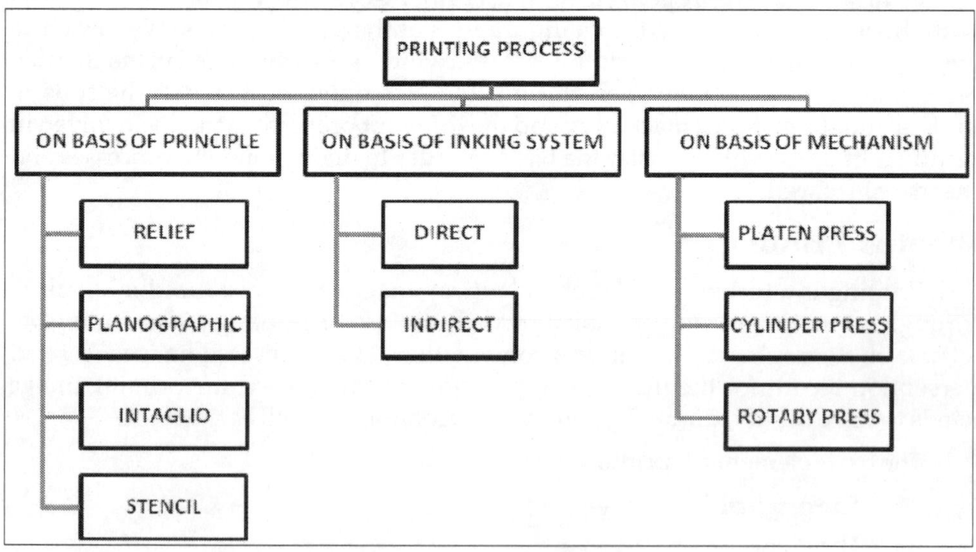

All the above given essentials can be a subject on their own, but we at preset are learning the basics of printing so we stick to the point.

How can we classify Printing?

Printing can be classified as follows:

As we can see there are 3 main categories by which we can classify the Printing processes:

- On the Basis of Principle
- On the Basis of Inking System
- On the Basis of Mechanism

Firstly we shall discuss the Principles of the Printing Processes.

Before we start, we must understand the terms Image area and Non-image area Image Area: is the part on the plate which will come as the printed data on the substrate. Please note that the term refers to all the data that to be printed on the substrate from the plate. The Data can be text also, it is not necessarily and image. Non-Image Area: is that part of the plate which does not contain the data for printing. Now we can start understanding the Principles that are used for the printing processes.

1. The Relief Process: In this process the image area stands out in 'relief' (*i.e.* raised above the surface), hence, leaving the non-image area 'recessed' or

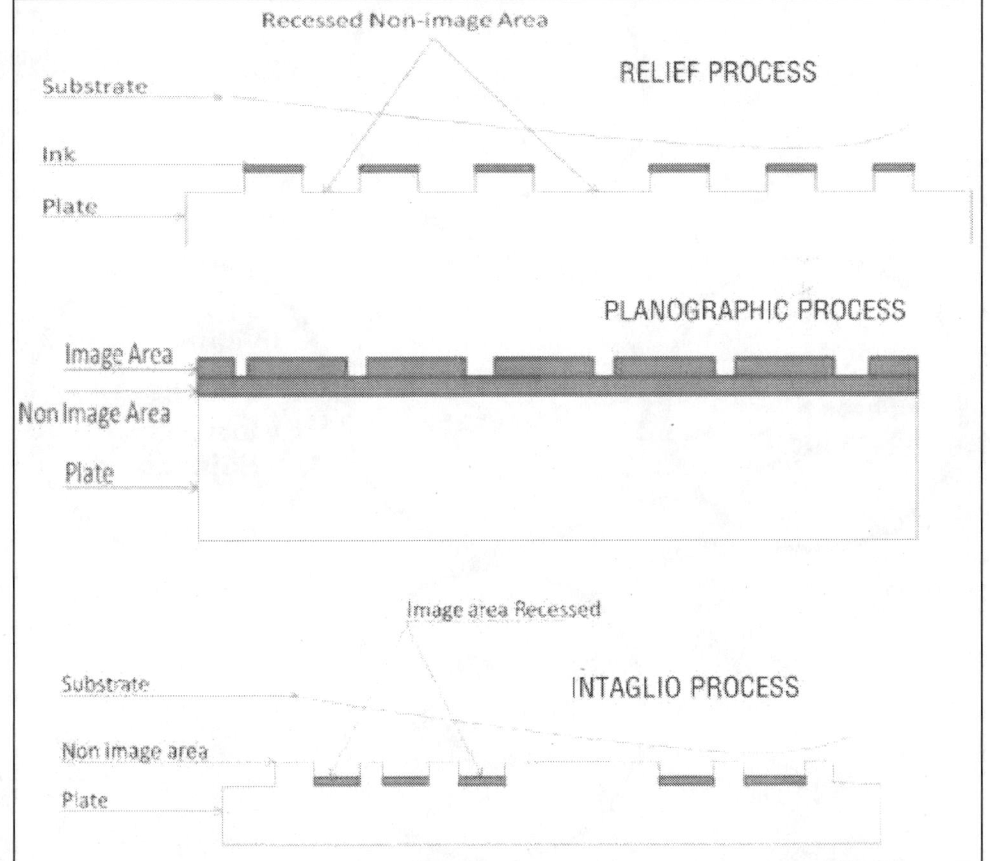

'broken' or discontinuous. The face of the plate is 'rolled' or coated with ink and a substrate is pressed on to the face of the plate, thereby giving the required data printed on the substrate. Hence the image alone from the surface of the plate is transferred onto the substrate.

2. The Plano-graphic Process: "Plane-O-Graphy" as the name suggests is printing with a flat surface. Here, the image and the non-image area seem to be on the same plane. This can be achieved due to the fact that oil and water do not mix physically, and secondly, any liquid has the tendency to penetrate into a porous and absorbent body. The printing area is essentially flat and the image is formed by the differential wetting of various plate areas by greasy lithographic ink and water. If I was to explain the system in 2 sentences I would say "Image Areas are Oleophilic, Non Image areas are Hydrophilic"
3. The Intaglio Process: Intaglio (Pronounced "in-tal-yo") is derived from the Italian word Intiaglieri meaning to engrave, cut out, or carve. In this process of printing, the ink carrying image area is recessed or broken or depressed

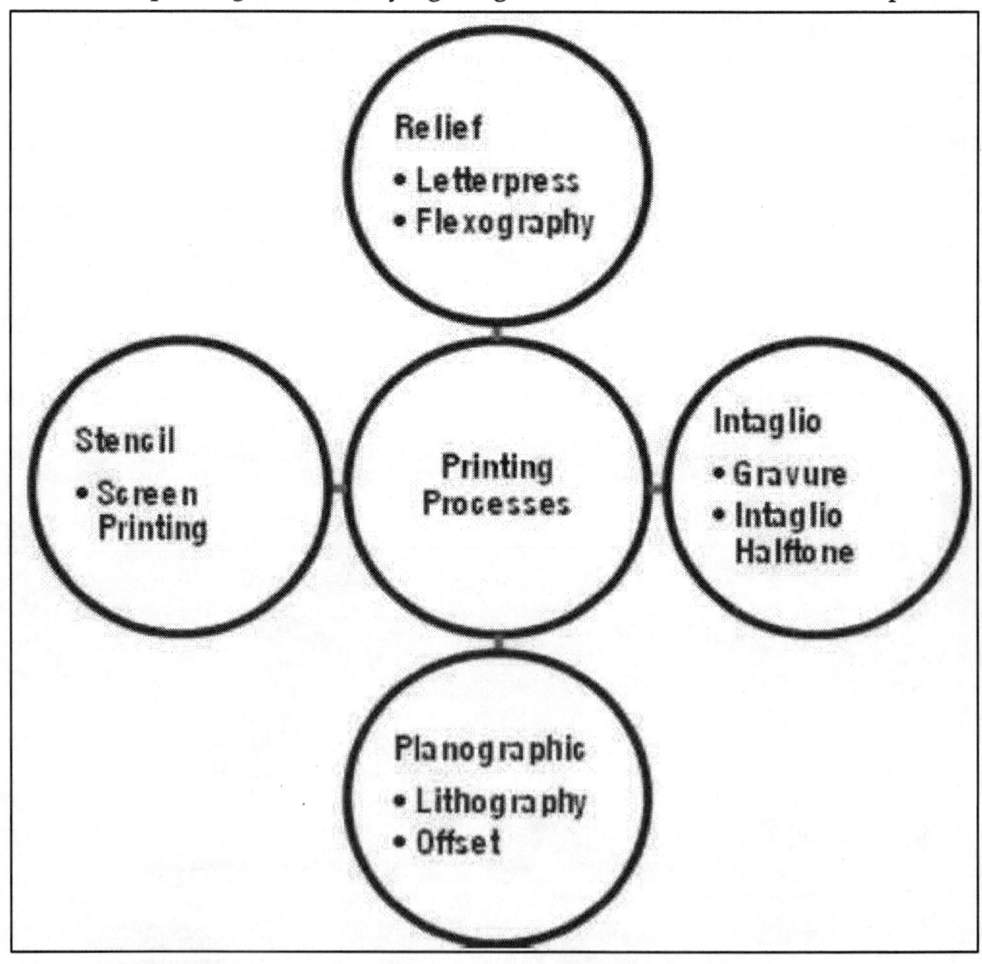

below the non image area. Although the surface is clean, ink remains in the engraved lines and will print when the paper is pressed against the plate. The whole plate is flooded and doctor blade is used to wash off the excess ink from the surface.

4. The Silk Screen or Stencil Printing: In this process the image carrier in not the plate, but is a woven material may be of cotton, nylon, silk or fine wire mesh through which a semi liquid paint or ink can pass through without clogging the mesh opening. The word silk was applied because of the usage of a screen made of silk threads. The principle is quite simple. The image area is open mesh and the non image area is blocked out, suitably, to prevent ink or paint seepage. It is used in packaging, advertising, and decoration of glass, on folding cartons, box wraps, sleeves, display boxes and gift wraps.

Now we shall discuss in short the inking systems.

1. Direct system: In this system the plate comes in direct contact with the substrate. Hence the ink is directly applied to the printing medium.
2. Indirect System: The ink is transferred onto an intermediate medium before it comes in contact with the substrate. Coming to the mechanisms of printing machinery, there are three types of machines:
   - Platen Press: The platen press prints by applying pressure for printing with a flat platen onto a printing plate installed on a flat plane. It is the oldest configuration.
   - Cylindrical Press: The Cylindrical Press prints by pressing a rotating impression cylinder onto a printing plate installed on a flat plane
   - Rotary Press: The Rotary press prints by bringing an impression cylinder into contact with a plate cylinder and rotating the two cylinders simultaneously

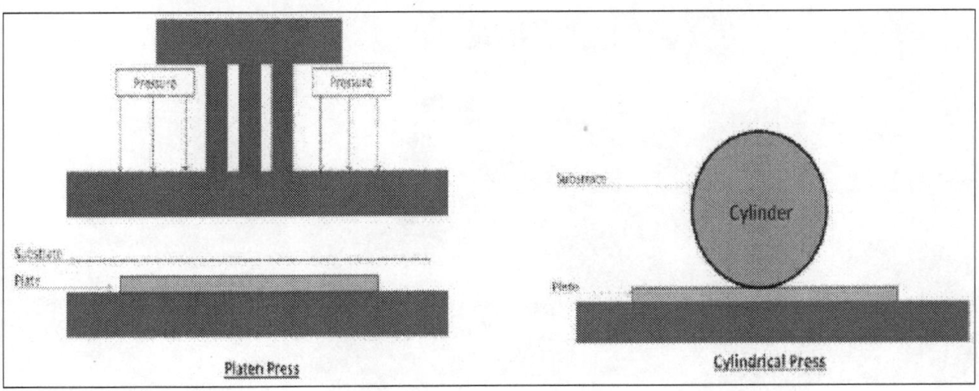

Platen Press Cylindrical Press

That now completes with the basic classification of printing.

What are the Sections of a Printing Press?

There are three main sections of any typical printing press, namely:

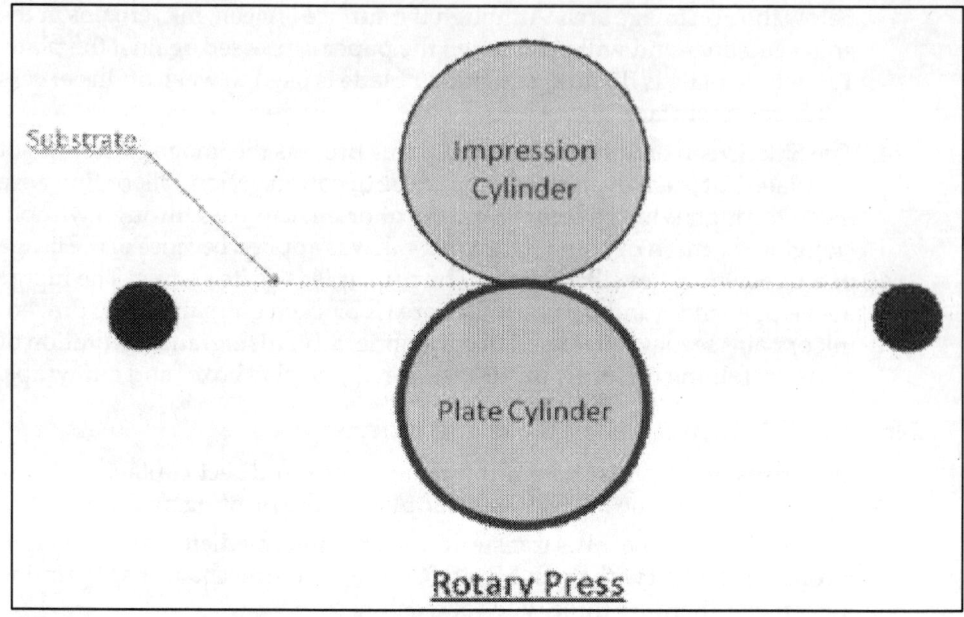

- Pre-Press Section
- Printing Section
- Post – Press Section

19

Testing of Filled Package

Functional Requirements

The basic function of a package is to protect and preserve the contents during transit from the manufacturer to the consumer. Protection is required against spillage, dirt, ingress and egress of moisture, insect infestation, contamination by foreign material, tampering, pilferage, etc. A package should preserve the contents in 'factory-fresh' condition during the period of storage and voyage by ensuring freedom from bacteriological attack, chemical reaction, etc. An important distinction is to be made here between two types of packaging:

Transit Packaging and Consumer Packaging

Transit Packaging

Most of the products entering the export trade require an outer package of one kind or another. An ideal transport package is one which offers protection against loss and damage during handling, transportation and storage and is at the same time, light and economical. The choice of the package, whether a wooden crate, a fibre board box or any other type – is to be governed by the characteristics of the product on the one hand and on the other, the kind of handling and transportation hazards likely to be encountered at various stages, not only in the exporting country but also during transit in the importing country.

A transit package is mainly expected to offer protection against handling and transportation hazards while a consumer that the product reaches the ultimate consumer in prime condition, free from contamination by dirt, moisture, heat, insects, etc. during transit and storage.

The best package for any particular purpose is the one which would protect the contents against the hazards the package would undergo during its journey at the minimum cost. The simplest and the most efficient way of testing packages is to carry out field trials with sufficient number of packages under the actual conditions of usage. However, such field trials are time consuming and require a fairly large number of packages sent out on several occasions by different routes. Evaluation of package performance or package testing is a means of shortening this process and of obtaining results in a shorter period with a reasonable degree of accuracy.

There are four main hazards of transport:

1. Drops and Impacts
2. Compression forces
3. Vibration and Vibration under stacking loads
4. Climatic variations

Equipments are available for testing packages for all these hazards.

Mechanical Tests

Drop Test

This test helps to measure the ability of the container and inside packing materials to provide protection to its contents, and to measure the ability of the container information useful improving the design of the container.

The common types of apparatus are available, *viz.* the Divided table top Drop test apparatus and the Hoist type with suitable slings, tripping device and hooks. In the first type of Tester, (which is very useful with packages which are difficult to sling) called the Table drop Tester, the package is held or supported in the desired position of fall and the trap door opened, thereby projecting the package on to the floor. The height of drop. The position of fall and the type of floor can be altered at will. The second type of tester consists of some form of release mechanism from which the package is suspended by means of a sling which allows it to be dropped in any particular position onto any type of floor from any selected height. The drop test may be divided into two procedures. Procedure – A shall be used for measuring the ability of the container to provide protection to its contents. In this procedure the drop test included cornerwise, edgewise and flat-wise drops. Procedure-B shall be used for measuring the ability of the container to with stand rough handling. In this procedure the drop test generally consists of cornerwise drops. The drop test results help to conclude the adequacy of packing and to ascertain the relative ability of the container to with stand the normal hazards encountered by a container from the time it is packed to the time it is opened for use.

The information which may be obtained from this test is of great value to the shipper and the container manufacturer. It is frequently possible, through the analysis of results to reduce the cost of package, or to improve the design of the package to prelude damage in transit. It affords the manufacturer of product a knowledge that the design of the product either does or does not, lend itself to economical packaging.

Vibration Test

The test is conducted to determine:

(i) The ability of the container

(ii) The protection offered by materials used for interior packing, and

(iii) The strength of the closures used

Vibration shocks could be encountered both during inplant handling and in shipment. Internal handling on conveyers and industrial trucks could produce damage to such highly fragile products as electronic components. However, the vast majority of vibration damage occurs in transit.

The effects of vibration are:

a) Structural failure in packaging materials through vibration

b) Reduction in container rigidity through weaving and swaying usually aggravated by wracking and skewing if the container bears a top load.

c) Exterior and interior damage through abrasion and

d) Damage to contents

A vibrating table is used for studying this type of hazard. The vibrating table consists of a bed which is driven by two eccentrics, one at each end, connected in phase with one another. To the top of the vibrating bed a platform is attached, and the platform describes a circular harmonic type of vibration when the equipment is running. The amplitude of the vibration on this instrument is fixed to one inch, and the frequency may be varied continuously from about 120 cycles a minute to about 360 cycles a minute. This test helps to study the possible package failure of content damage, and the weakness (loosening of bonds, separation of container layer, abrasion of smooth surfaces, loss of resilience in cushioning material etc.) that may develop during transit.

Compression Test

This test is carried out, generally, on empty containers to measure the ability of the container to resist external compressive loads applied to faces and when applied to diagonally opposite edges or corners. Compression stresses may be caused by (1) static condition due to the superimposition of weight onto merchandise of by (2) dynamic stresses of impact stresses. While the static impression is primarily experienced in stacking, the dynamic compressive stresses generally result from impact shock in handling and transit. During storage or transport, packages are stacked to the maximum height possible. Consequently the bottom layer is subjected to compression and hence has to bear the load of the packages stacked over it. If the container is not strong enough the force of compression is transmitted to the contents. This is particularly so, in case of fibre board containers and the currently used bamboo baskets for fruits and vegetables. Thus, the container used should have adequate compression strength for a given purpose. This is determined by means of standard compression tester. The data obtained would indicate the strength of the container for short period of stacking, for long period it is preferable to take the compression strength as one third of the test value. For dynamic conditions the performance strength could be taken as the test value multiplied by 3/2 in the vertical direction.

Inclined Impact Test (Contour Test)

This test helps to study the extent of crushing, breaking, cracking, distortion and shifting during handling storage and transport. These are mainly due to shunting shocks. This test is carried out to determine the ability of a container to withstand impact stresses and also to determine the ability of a container or interior packing, or both, to provide protection to the contents when subjected to impact stresses. This test is particularly valuable for testing large, odd shaped, or heavily loaded containers that would be difficult or impossible to test by other methods.

The study is carried out generally with an Inclined-plane Impact Tester. This consists of a track inclined at 10 per cent to the horizontal, on which a dolly can be released so as to impact at the wooden buffers placed at right angles across the end. According to the distance of the incline from which the dolly is released, the impact speed can be varied upto about 8 m.p.h. The results of the tests, are indicative of the effects experienced by the package when it is subjected to shunting shocks of side impacts during transit. From the results obtained, it is possible to design suitable containers or to improve the design of the existing containers or interior packing.

Rolling Test

The rolling test is performed as follows:

The package, standing on its base, is slowly pushed to rotate about one edge of the base and allowed to fall over on the side face. After this impact, the package is lifted, balanced onn its edge and then allowed to fall over on the top. The procedure is repeated until the package once again stands on its base. The whole sequence from the base to base traveling in the direction at a right angle to that taken in the first half of the operation is repeated. This test helps to evaluate the overall strength of the container and the cushioning provided inside, and any failure of the contents. Results obtained from this test help to design a suitable container, assess the adequacy of the cushioning material and to arrive at suitable and economical placement of the contents in the bulk container.

Drum Test

Drum test helps to evaluate loaded shipping containers with respect to general overall durability and for the protection afforded to the contents against certain hazards of handling and shipments (This method is suitable for testing boxes or crates made of metal, wood fibreboard or combinations of these materials).

Climatic Test

1. Rain Test

A simulated rainfall of 4x1 inches per hour should be provided by a water spray nozzle designed to emit water in small droplets rather than a fine mist. The rainfall should be dispersed uniformly over the test area. Duration of the test is generally two hours.

2. Sand and Dust Tests

The purpose of sand and dust tests is to evaluate the resistance of a package to the penetration of sand and dust, to determine the erosive effects of blowing sand and dust on exposed items and to determine the immediate effects such as the malfunction of moving parts, by the presence of sand and dust. A standardized mixture of sand and dust of density 0.1 to 0.5 gm / cu. Ft. is used to create an atmosphere for this. The temperature of this atmosphere is maintained at 77oF for a period of 6 hours and then increased to 160oF for another 16 hours. A standard sand and dust velocity of 200′ per min. is maintained throughout the test unless an optional velocity of 2,300 ft. per min. is specified. Items which are always shielded from blowing sand and dust should be tested under standard velocity conditions, 200′ per min. Items which may be exposed to blowing sand and dust should be tested under optional velocity conditions 2,300 per min.

3. Salt Spray Test

The purpose of the salt spray test is to evaluate the resistance of the package to corrosion by salt spray and to serve as a general standard for corrosion resistance. No correction exists between the test time and the natural exposure time. The package is exposed for 50 hours, to a wet, dense fog generated by the automation of a 20 per cent water solution of sodium chloride. The solution shall be maintained at a pH of between 6.5 to 7.2, the temperature of the fog being maintained at 95oF.

4. Fungus Resistance Tests:

All materials used in the fabrication of shipping containers should be tested for fungus resistance. A composite spore suspension is prepared as follows: 10 ml of distilled water containin approximately 0.005 per cent dioctyl sodium sulfacuccinate is introduced directly into each culture. The culture is then raked with a sterile wire and agitated to insure a well sporated suspension. The test item is then placed in a humidity of 25 per cent and a temperature of 86oF for 28 days. If, at the end of 28 days, 25 per cent or less of the samples or test items have fungus growth on more than 4 per cent of their exposed surface, the material is considered to be fungus nutrient. If 2 per cent or less of the exposed surface is covered, the material is considered to be inert for fungus growth.

The foregoing gives only a brief idea of the possible laboratory and field tests. These tests give the following correlations to obtain a satisfactory picture on package performance.

- A correlation between the performance of the package in service and the laboratory transport test which reproduces in some degree the events which would happen to the package in the field. Such a laboratory transport test is capable of evaluating a package as to its probable efficiency. The user of the package is primarily concerned with this correlation.
- A correlation between the laboratory transport test on the package and tests on the empty container and any other component such as cushions etc. thus would enable the package maker to use simpler control tests on bulk production.

- The third correlation to be made between the materials of construction and the containers.

The particular type of testing used in any instance will be dependent on the circumstances and the information desired.

In general, it might be said, that all these tests would help to conclude:

- The overall durability of the container
- Any possible reduction in the cost of packaging
- Improvement required in the design of the box to preclude damage in transit
- Adequacy of the cushioning used
- In short, a series of tests would enable to determine the transportworthiness of the package.

20

Packaging Laws and Regulations

The link between food packaging and consumer protection is of high significance. A package is a vehicle of safety and achieves the objective of delivering safe, wholesome, nutritious food to the consumer. To safeguard the interests of the consumer and the society at large, Packaging Laws and Regulations have been introduced by the Government. The Indian Regulatory System falls under the category of compulsory legislations formulated by the various ministries and voluntary standards framed by various organisations to serve the country. The National Regulatory System is shown in Table 1.

The Packaging Laws and Regulations for food products are mainly covered under:

- The Standards of Weights and Measures Act, 1976 and the Standards of Weights and Measures (Packaged Commodities) Rules, 1977 (SWMA).
- The Prevention of Food Adulteration Act, 1954 and the Prevention of Food Adulteration Rules, 1955 and its first ammendment, 2003 (PFA).
- The Fruit Products Order, 1955 (FPO)
- The Meat Food Products Order, 1973 (MFPO)
- The Edible Oil Packaging Order, 1998
- The Agmark Rules

The Standards of Weights and Measures Act (SWMA)

Till about 25 years ago or so, the consumer was not sure that he was getting for his money, the right weight or volume of the packaged products. Things were pretty

TABLE 1

FOOD LAWS/REGULATIONS AND MINISTRIES INVOLVED

REGULATIONS	MINISTRY OF CONSUMER AFFAIRS	MINISTRY OF FOOD PROCESSING INDUSTRIES	MINISTRY OF AGRICULTURE	MINISTRY OF HEALTH AND FAMILY WELFARE	MINISTRY OF COMMERCE	MINISTRY OF FOOD AND CIVIL SUPPLIES
Essential Commodities Act, 1955	Solvent extracted oils, De-oiled Meal and Edible Flour Control Order 1967. Mandatory inspection	Fruits Product Order 1955 Mandatory inspection	Meat Food Product Order 1973 Mandatory inspection			
Standards of Weight Measures Act, 1976	SWMA rules, 1977. Packed foodstuffs must adhere to quality declaration					
Agricultural Produce (Grading and Marketing) Act 1937			Agmark standard for raw and semiprocessed products. Voluntary inspection			
Prevention of Food Adulteration Act 1954				Protects consumer against inferior quality and adulteration		
Codex Standard (CAC) 1964 (not a law)				Endorsement by WTO under SPS and TBT. De-facto mandatory		
Export (Quality Control and Inspection) Act, 1963					Pre-shipment inspection	
Bureau of Indian Standards, 1986						HACCP 9000 certification. Voluntary inspection

The Act also specifies the base units for:

• Length – Metre , • Mass – Kilogram, • Time – Second , • Electric Current – Ampere

• Thermodynamic Temperature – Kelvin

• Luminous Intensity – Candela

• Base Unit of Numeration – International form of Indian numerals

chaotic with different states having their own system of weights and measures. Adding to the confusion was the common practice of putting the same quantity of products in packs of different sizes, some containers being half-filled, and, as if this was not enough, quantities of contents on the packs were not stated in terms of units of weight or measure but declared as "family pack", "economy size", "full size" and so on. This unhappy state of affairs prevailed till 1976 when the Government of India, brought forward a wise and enlightened piece of legislation in the form of The Standards of Weights and Measures Act, 1976 (SWMA). Some of the important aspects of SWMA are highlighted here.

Standard Units (Section 4)

Chapter I of the Act is on establishment of standard units. This chapter clearly tell us the units of weights and measures to be followed. It states that every unit shall be based on the metric system. For this purpose, the units to be adopted are the International System of units recommended by the General Conference on Weights and Measures and such additional

Declaration on Packaged Commodities for Intrerstate Trade or Commerce

In Chapter IV (section 39), the Act stipulates that for interstate trade or commerce of commodities in packaged form, intended to be sold or distributed, every commodity in packaged form has to bear upon it, on a label securely attached to it, a definite, plain and conspicuous declaration of:

- Identity of the commodity in the package
- Net quantity, in terms of the standard unit of weight or measure, of the commodity in the package
- Where the commodity is packaged or sold by number, the accurate number of commodity contained in the package
- The unit sale price of the commodity in the package
- The sale price of the package

Further requirements include:

- Every package should bear the name of the manufacturer and also of the packer or distributor.
- The statement as to the net weight, measurement or number of the contents should not have any expressions, which tend to qualify such weight, measurement or number. (Exceptions to this are commodities which may undergo changes in weight or measure due to climatic variations; examples – bread, soap, etc. where the qualifying statement "when packed" may be added to the net weight or measure).
- Where there is undue proliferation of weight, measure or number in which any commodity is being sold and such undue proliferation impairs, in the opinion of the Government, the reasonable ability of the consumers to make a comparative assessment of the prices after considering the net quantity or number of such commodity, the Government may prescribe standard quantities or numbers for any commodity.

- Where the retail price of a commodity is stated in any advertisement, the net quantity or number of the commodity must be conspicuously declared in the advertisement along with the price.
- A package containing a commodity, which is filled less than the prescribed capacity of such package cannot be sold or distributed except where it is proved that the package is so filled with a view to (a) giving protection to the contents of the package or (b) meeting the requirements of machines used for enclosing the contents of such packages.
- The Central Government may, by rules, specify reasonable variations in the net contents of the commodity in a package as may be caused by the method of packing or the ordinary exposure which may be undergone by the commodity after it has been introduced in the market place.

This very comprehensive and far-reaching Act has put an end to the state of near anarchy in the trading of packaged goods. The clearly specified requirements in the Act have also provided a challenge to packaging development experts and label copy specialists who have to include statutory and promotional copy in the limited space available on labels and on packages themselves. However irksome they may appear, the provisions of this Act are welcome because they offer to the consumer a measure of protection which is not so apparent in many other legal requirements.

Standard Packages

Under the Standards of Weights and Measures (Packaged Commodities) Rules, rules have been framed specifying provisions for the retail sale of packaged goods. One of the most important rules is with respect to the requirements that specific commodities are to be packed and sold only in standard packages. As per the Third Schedule, food products and their respective package capacities are given in Table 2.

Maximum Permissible Error

In reference to the same rules as above, under the First Schedule, maximum permissible errors in relation to the quantity contained in individual packages is specified as given in Table 3 for food packages. Table 4 gives the maximum permissible errors in relation to net quantities of packaged commodities (food) not specified in the First Schedule.

As per the Fifth Schedule of the SWMA Rules, commodities to be sold by weight, measure or number are indicated. Table 5 gives the details of the same with respect to food products.

Label Declarations

In the SWMA Rules, the declaration to be made on every retail package has been detailed. The declarations are to be made with respect to the following:

- The name and address of the manufacturer or where the manufacturer is not the packer, the name and address of the manufacturer and packer.
- The common or generic names of the commodity contained in the package.

TABLE 2

Commodities to be Packed in Specified Quantities (Standard Packages) as per The Third Schedule of SWMA Rules

COMMODITIES	QUANTITIES IN WHICH TO BE PACKED
Baby food	200g, 500g, 1 kg, 2 kg, 5 kg and 10 kg – Any manufacturer or packer packing baby food in 400g and weaning food in 500g shall not be allowed to do so beyond 30.6.95
Weaning food	200g, 400g, 1 kg, 2 kg, 5 kg and 10 kg – Publication of this notification in the official gazette
Biscuits	25g, 50g, 75g, 100g, 150g, 200g, 250g, 300g and thereafter in multiples of 100g up to 1kg
Bread including brown bread but excluding bun	100g and thereafter in multiples of 100g
Uncanned packages of butter and margarine	25g, 50g, 100g, 200g, 500g, 1 kg, 5 kg and thereafter in multiples of 5 kg
Cereals and pulses	100g, 200g, 500g, 1 kg, 2 kg, 5 kg and thereafter in multiples of 5 kg
Coffee	25g, 50g, 100g, 200g, 500g, 1 kg and thereafter in multiples of 1 kg
Tea	25g, 50g, 100g, 200g, 500g, 1 kg and thereafter in multiples of 1 kg
Materials which may be reconstituted as beverages	25g, 50g, 100g, 200g, 500g, 1 kg and thereafter in multiples of 1 kg
Edible oils, vanaspati, Ghee, butter oil	50g, 100g, 200g, 500g, 1kg, 2kg, 3kg, 5kg, and thereafter in multiples of 5kg. If net quantity is declared by volume the same number in millilitres or litres, as the case may be. If the net quantity is declared by volume then the equivalent quantity in terms of mass to be declared in brackets, in same sizes of letters/numerals
Milk Powder	Below 50g no restriction, 50g, 100g, 200g, 500g, 1 kg and thereafter in multiples of 500g
Rice (powdered), flour, atta, rawa and suji of 5 kg	100g, 200g, 500g, 1 kg, 2 kg, 5 kg and thereafter in multiples5 kg
Salt	Below 50g. in multiples of 10g; 50g, 100g, 200g, 500g, 750gms, 1 kg, 2 kg, and thereafter in multiples of 5 kg
Aerated soft drinks and non-alcoholic beverages	100ml, 150ml, 200ml, 250ml, 300ml, 330ml (in cans only), 500ml, 750ml, 1 litre, 1.5 litre, 2 litre, 3 litre, 4 litre, and 5 litre
Mineral water and drinking water	100ml, 130ml, 150ml, 200ml, 250ml, 300ml, 330ml, 500ml, 600ml, 750ml, 1 litre, 1.2 litre, 1.5 litre, 2 litre, 3 litre, 4 litre, and 5 litre. The sizes 130ml, 330ml, 600ml and 1.2 litre shall be allowed only for a period of 3 years from the date of notification. (26th Nov. 2001)

- The net quantity in terms of the standard unit of weight or measure, of the commodity contained in the package or where the commodity is packed or sold by number, the number of commodity contained in the package.
- The month and year in which the commodity is manufactured or pre-packed. (Provided that for Packages containing food articles, the provisions of the Prevention of Food Adulteration Act (PFA), 1954 (37 of 1954) and the rules made thereunder shall apply).
- The retail price of the package.
- The retail sale price of the package.

TABLE 3

Maximum Permissible Error in Relation to Quantity Contained in Individual Package as per the First Schedule of SWMA Rules

Description of Commodity	Quantity Declared	Maximum Permissible Error
Biscuits	(i) Up to and equal to 500g (ii) Above 500g	7.0% 6.0%
Bread	(i) Up to and equal to 400g (ii) Above 400g up to and equal to 800g (iii) Above 800g up to and equal to 1200g	8.0% 6.0% 4.0%
Ghee, vanaspati and edible oil	(i) Up to and equal to 1 kg/litre (ii) Above 1kg / litre up to and equal to 2kg / litre. (iii) Above 2 kg / litre up to and equal to 4kg / litre. (iv) Above 4kg / litre	2.0% 1.5% 1.25% 0.6%
Infant food including malted milk food	(i) Up to and equal to 100g (ii) Above 100g up to and equal to 1 kg (iii) Above 1 kg	5.0% 4.0% 3.0%
Liquid milk	(i) Up to and equal to 100ml (ii) Above 100 ml up to and equal to 250ml (iii) Above 250 ml	5 ml 8 ml 10 ml
Provisions sold in polythene bags or plastic bags, food grains, pulses, edible seeds, spices(whole or broken but not powdered), powderedcommodities, (such as, chilli powder,pepper powder, coffee powder, washing soda, atta, table salt and thelike), dry fruits, seeds and other commodities (such as, sugar gur, khandsari and the like)	(i) Up to and equal to 100g (ii) Above 100g up to and equal to 500g (iii) Above 500g up to and equal to 1 kg (iv) Above 1 kg	3.0% 2.0% 1.5% 0.75%
Tea	For all quantities	2.0%

The maximum permissible error specified as percentage shall be rounded off to the nearest one-tenth of a g or ml, of declared quantities less than or equal to 1000g or ml and to the next whole g or ml for declared quantities above 1000g or ml.

TABLE 4

Maximum Permissible Errors on Net Quantities Declaredby Weight or by Volume (not specified in First Schedule)as per SWMA Rules (Second Schedule)

Declared Quantity (g or ml)	Maximum Permissible Error in Excess or in Deficiency	
	As Percentage of Declared Quantity	g or ml
Up to 50	9	–
50 to 100	–	4.5
100 to 200	4.5	–
200 to 300	–	9
300 to 500	3	–
500 to 1000	–	15
1000 to 10000	1.5	–
10000 to 15000	–	150
More than 15000	1.0	–

TABLE 5
Commodities to be Sold by Weight, Measure or Number as per The Fifth Schedule of SWMA

COMMODITY	WHETHER DECLARATION TO BE EXPRESSED IN TERMS OF WEIGHT MEASURES OR NUMBER OR TWO OR MORE OF THEM
Curd	Weight
Fruits, all kinds	Number or weight
Edible oil, vanaspati, ghee and butter oil	Weight or volume
Honey, malt extract, golden syrup treacle	Weight
Ice cream and other similar frozen products	Weight or volume
Rasgulla, Gulab Jamun and other sweet preparations	Weight
Sauce, all kinds	Weight

Rules

Where any package material bearing thereon the month in which any commodity was expected to have been pre-packed is not exhausted during that month, such packaging material may be used for pre-packing the concerned commodity produced or manufactured during the next succeeding month and not thereafter, but the Central Government may if it is satisfied that such packaging material could not be exhausted during the period aforesaid by reason of any circumstance beyond the control of the manufacturer or packer, as the case maybe, extend the time during which such packaging material may be used, and where any such packaging material is exhausted before the expiry of the month indicated thereon, the packaging material intended, to be used during the next succeeding month may be used for pre-packing the concerned commodity; provided that the said provision shall not apply to the packages containing food products, where the "Best before or Use before" period is ninety days or less from the date of manufacture or packing. General Provisions Relating to Declaration of Quantity

1) In declaring the net quantity of the commodity contained in a package, the weight of wrappers and materials other than the commodity shall be excluded; provided that where a package contains a large number of small items of confectionery, each of which is separately wrapped, the net weight declared on the package containing such confectionery or on the label thereof may include the weight of such immediate wrappers, if and only if, the total weight of such immediate wrappers does not exceed:
   (i) 8 per cent, where such immediate wrapper is a waxed paper or any other paper, with wax or aluminium foil (under strip), or
   (ii) 6 per cent, in case of any other paper, of the total net weight of all the items of confectionery contained in the package minus the weight of immediate wrapper.
2) Where a commodity in a package is not likely to undergo any variation in weight or measure, on account of the environmental conditions, the quantity declared on the package shall correspond to the net quantity, which will be received by the consumer, and the declaration of quantity on such package shall not be qualified by the words "when packed" or the like.

3) Save as otherwise provided in sub-rule (4), where a commodity in package is likely to undergo variations in weight or measure on account of environmental conditions and such variation is negligible, the declaration of quantity in relation to such package shall be made after taking into account such variation so that the consumer may receive not less than the net quantity of the commodity as declared on the package, and the declaration of quantity on such package shall not also be qualified by the words "when packed" or the like.
4) The declaration of quantity in relation to commodities specified in the Fourth Schedule, that is to say, commodities which are likely to undergo significant variation in weight or measures on account of environmental or other conditions may be qualified by the words "when packed".

Symbols for Unit

The symbols for International System of units and none other shall be used in furnishing the net quantity of the package.

Illustrations:

Kilogram	Kg
Gram	g
Milligram	mg
Litre	l
Millilitre	ml
Metre	m
Centimetre	cm
Millimetre	mm
Squaremetre	m^2
Square centimeter	cm^2
Cubic metre	m^3
Cubic centimetre	cm^3

Symbols shall not be given in capital form except for the unit derived from a proper name, period *i.e.* a dot after symbols shall not be put. As far as possible symbols shall always be written in the singular form, *i.e.*'s' shall not be added.

General Guidelines on Giving Declarations

As far as possible, all declarations required to be made under SWMA Rules should appear on the principal display panel. The principal display panel is defined as that part of the package that is intended, or likely to be displayed, presented or shown or examined by the consumer under normal and customary conditions of display, sale or purchase of the commodity contained in the package. Every declaration which is required to be made on a package should be legible, prominent, definite, plain and unambiguous and should be given in a specified minimum size as given in Tables 6 and 7, depending on the area of the principal display panel. Specific guidelines are given for computing the area of the principal display panel.

TABLE 6

Minimum Height of Numerals

Net Quantity in Weight / Volume	Minimum Height in mm	
	Normal Case	When Blown Formed, Moulded, Embossed or Perforated on Container
Up to 200 g/ml	1	2
Above 200 g/ml up to 500 g/ml	2	4
Above 500 g/ml	4	6

TABLE 7

Minimum Height of Numerals

Net Quantity in Length Area or Number, Area of Principal Display Panel	Minimum Height in mm	
	Normal Case	When Blown Formed, Moulded, Embossed or Perforated on Container
Up to 100 cm square	1	2
Above 100 cm square up to 500 cm square	2	4
Above 500 cm square up to 2500 cm square	4	6
Above 2500 cm square	6	6

Violation of Law

What happens if the law is violated? To explore this, let us move to part VI of the Act. This part provides penalty for different offences. The penalty for violation of Section 39 is in Section 63. If any person packs, distributes, stores, delivers or sells commodities, which does not meet the requirements of the Act and the Packaged Commodities Rules, can be punished by a fine which may extend up to Rs.5000. If the offence is repeated, the penalty can be imprisonment of up to five years. Section 72 provides for prosecution before a magistrate. But in the first instance, the department tries to settle the case with the offender. This is called compounding. The provision on compounding is contained in Section 74. The authorised officer of the department of Legal Metrology can compound a case, with the consent of the offender, by charging a compounding fee. This fee can be up to Rs. 5000. For the next three years, a subsequent offence cannot be compounded, it will have to be taken to the court. But after three years, an offender again becomes eligible to get a case compounded. Similarly, Section 74 provides for offences by companies and other body corporate. Both the persons, master and the servant, are jointly responsible. Thus, when a firm commits an offence, the company and the person who is the cause for commission of the offence, are both jointly responsible.

The Prevention of Food Adulteration Act

Food is one of the basic necessities for sustenance of life. Pure fresh and healthy diet is most essential for the health of the people. It is not wrong to say that community health is national wealth. dulteration of food was so rampant, widespread and persistent that nothing short of a somewhat drastic remedy in the form of a comprehensive legislation became the need of the hour.

To check this sort of anti-social evil, a concerted and determined onslaught was launched by the Government by introduction of the Prevention of Food Adulteration Bill in the Parliament to herald an era of much needed hope and relief for the consumers at large. The Prevention of Food Adulteration Act and Rules have provided standards for a large variety of food. Unfortunately, the importance on packaging is not adequately reflected except in a few cases such as infant food items, drinking water. The responsibility of adequate packaging of food and its safety falls on the manufacturer of the food product. The Prevention of Food Adulteration Act, 1954 (PFA) prohibits manufacture, storage and sale of adulterated food. The violation of law is prosecuted before a magistrate's court. The punishment is mandatory imprisonment for a minimum of three months. No surprise, the PFA is dreaded by the food industry. To understand the PFA, one must know what are the items included in the category of food and what the law considers to be adulteration of food.

Food and Adulteration

As per PFA, Food includes everything, which is consumed by human beings or even used for preparing items of human consumption. Thus cereals, oil, sugar, cooked food, drinks, spices, colouring matters, flavouring matter etc. are all included in the category of food. It excludes water and drugs. However, packaged natural water and packaged mineral watern are considered to be food.

We ordinarily mean by adulteration to "debase, falsify by mixing with something inferior or spurious". By adulterated food, people also mean rotten, putrefied, insect infested or poisonous food. As per the Act, a food is deemed to be adulterated:

(a) If the article sold by a vendor is not of the nature, substance or quality demanded by the purchaser and is to his prejudice, or is not of the nature, substance or quality which it purports or is represented to be.

(b) If the article contains any other substance, which affects, or if the article is so processed as to affect, injuriously the nature, substance or quality thereof.

(c) If any inferior or cheaper substance has been substituted wholly or in part for the article so as to affect injuriously the nature, substance or quality thereof.

(d) If any constituent of the article has been wholly or in part abstracted so as to affect injuriously the nature, substance or quality thereof.

(e) If the article had been prepared, packed or kept under insanitary conditions whereby it has become contaminated or injurious to health.

(f) If the article consists wholly or in part of any filthy, putrid, rotten, decomposed or diseased animal or vegetable substance or is insect-infested or is otherwise unfit for human consumption.

(g) If the article is obtained from a diseased animal.

(h) If the article contains any poisonous or other ingredient which renders it injurious to health.

(i) If the container of the article is composed, whether wholly or in part, of any poisonous or deleterious substance which renders its contents injurious to health.

(j) If any colouring matter other than prescribed in respect thereof is present in the article, or if the amount of the prescribed colouring matter, which is present in the article are not within the prescribed limits of variability.

(k) If the article contains any prohibited preservative or permitted preservative in excess of the prescribed limits.

(l) If the quality or purity of the article falls below the prescribed standard or its constituents are present in quantities not within the prescribed limits of variability, but, which renders it injurious to health.

(m) If the quality or purity of the article falls below the prescribed standard or its constituents are present in quantities not within the prescribed limits of variability, but which does not render it injurious to health.

Provided that, where the quality or purity of the article, being primary food, has fallen below the prescribed standards or its constituents are present in quantities not within the prescribed limits of variability in either case, solely due to natural causes and beyond the control of human agency, then, such article shall not be deemed to be adulterated within the meaning of this sub-clause.

Packaging and Storage Requirements

Accordingly, Part IX (Rule 49(5)) of the PFA Rule States: A utensil or container made of the following materials or metals, when used in the preparation, packaging and storing of food shall be deemed to render it unfit for human consumption:

- containers which are rusty
- enameled containers which have become chipped and rusty
- copper or brass containers which are not properly tinned
- containers made of aluminium not conforming in chemical composition to IS: 20 Specification for cast aluminium and aluminium alloy for utensils or IS: 21 specification for wrough aluminium and aluminium alloy for utensils
- Container made of plastic materials not conforming to the following Indian Standards Specification, used as appliances or receptacles for packing or storing, whether partly or wholly, food articles, namely:
    - IS: 10146 (Specification for polyethylene in contact with food stuffs)
    - IS: 10142 (Specification for styrene polymers in contact with foodstuffs)
    - IS: 10151 (Specification for Poly Vinyl Chloride (PVC) in contact with food stuffs)
    - IS: 10910 (Specification for polypropylene in contact with foodstuffs)
    - IS: 11434 (Specification for ionomer resins in contact with foodstuffs)
    - IS: 11704 (Specification for Ethylene Acrylic Acid (EAA) co-polymer)
    - IS: 12252 (Specification for Polyalkylene Terephathalates (PET))

- IS: 12247 (Specification for Nylon 6 polymer)
- IS: 13601 – Ethylene Vinyl Acetate (EVA)
- IS: 13576 – Ethylene Metha Acrylic Acid (EMAA)

- Tin and plastic containers once used shall not be re-used for packaging of edible oil and fats.

The PFA Rules also stipulate that certain food items such as confectionery (weighing more than 500 grams), protein rich atta, protein rich maida, blended edible vegetable oil, coloured and flavoured table margarine, fat spread, spices and condiments shall be sold in packed condition only.

Other Packaging Requirements under PFA

- For infant milk food, infant formula milk cereal based weaning food and processed cereal based weaning food, the rules state that: The product shall be packed in hermetically sealed, clean and sound containers or in flexible packs made from film or combination of any or substrate made of board paper, polyethylene, polyester metallised film or aluminium foil in such a way so as to protect it from deterioration.
- For meat and meat products, the product shall be packed in hermetically sealed containers and subjected to heat treatment followed by rapid cooling to ensure that the product is shelf-stable.

  The sealed container shall not show any change on incubation at 35°C for 10 days and 55°C for 5 days.
- For natural mineral water, naturally carbonated natural mineral water, and packaged drinking water, the rules stipulated regarding the packaging materials are:

  It shall be packed in clean, hygienic, colourless, transparent and tamperproof bottles/containers made of Polyethylene (PE) conforming to IS: 10146 or Poly Vinyl Chloride (PVC) conforming to IS: 10151 or Polyalkylene Terephthalate (PET and PBT) conforming to IS 12252 or Polypropylene conforming to IS: 10910 or food-grade Polycarbonate or sterile glass bottles suitable for preventing possible adulteration or contamination of the water. All packaging materials of plastic origin shall pass the prescribed overall migration and colour migration limits.

Declarations and Labeling

The other aspect of regulations under the PFA is with respect to declarations/labeling. Any packaged food, which does not conform to these requirements under the PFA is deemed "misbranded". As per the Act, an article of food shall be deemed to be misbranded:

- If it is an imitation of, or is a substitute for, or resembles in a manner likely to deceive, another article of food under the name of which it is sold, and is not plainly and conspicuously labeled so as to indicate its true character.

- If it is falsely stated to be the product of any place or country.
- If it is sold by a name which belongs to another article of food.
- If it is so coloured, flavoured or coated, powdered or polished, that the fact that the article is damaged, is concealed or if the article is made to appear better or of greater value than it really is.
- If false claims are made for it upon the label or otherwise.
- If, when sold in packages, which have been sealed or prepared by or at the instance of the manufacturer or producer and which bear his name and address, the contents of each packages are not conspicuously and correctly stated on the outside thereof within the limits of variability prescribed under this Act.
- If the package containing it, or the label on the package bears any statement, design or device regarding the ingredients or the substances contained therein, which is false or misleading in any material particular; or if the package is otherwise deceptive with respect to its contents.
- If the package containing it or the label on the package bears the name of a fictitious individual or company as the manufacturer or producer of the article.
- If it purports to be, or is represented as being, for special dietary uses, unless its label bears such information as may be prescribed concerning its vitamin, mineral, or other dietary properties in order to sufficiently inform its purchaser as to its value for such uses.
- If it contains any artificial flavouring, artificial colouring or chemical preservative, without a declaratory label stating that fact, or in contravention of the requirements of this Act or rules made there under.
- If it is not labeled in accordance with the requirements of this Act or rules made there under.

Part VII of the Rules deals with the Packing and Labeling of Food. As per these rules, the following are required:

- The name, trade name or description of food contained in the package.
- The names of ingredients used in the product in descending order of their composition by weight or volume as the case may be. If artificial flavouring is used, the chemical names of the flavour need not be declared, but, in the case of natural flavouring substances or nature-identical flavouring substances, the common name of the flavour is to be mentioned on the pack. If the food contains any ingredient in part or whole from animal origin (meat, fish, poultry eggs), a declaration is to be made by a symbol and a colour code stipulated for this purpose, to indicate the product as Non-vegetarian Food. The symbol should be on the principal display panel in close proximity to the name or rand name of the food as indicated in clause (16) of sub-rule (zzz) of rule 42. The symbol shall consist of a brown colour filled circle having a diameter not less than the minimum size specified in Table 8.

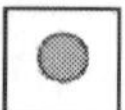

Non-vegetarian Food Vegetarian Food

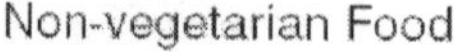

TABLE 8

Symbol Sizes

Area of Principal Display Panel	Minimum Size of Diameter in mm
Up to 100 cm square	3
Above 100 cm square up to 500 cm square	4
Above 500 cm square up to 2500 cm square	6
Above 2500 cm square	8

Similarly, for vegetarian food a similar symbol with green colour circle and square as indicated in clause (17) of sub-rule (zzz) of rule 42 will be displayed. The symbol shall be prominently displayed on the package having contrast background and in close proximity to the name or brand name of product, and also on the labels, pamphlets, leaflets, and advertisements in any media.

- The name and complete address of the manufacturer, or importer, or vendor or packer to be declared.
- A declaration is to be made for the net weight or number or measure of volume of content in the case of biscuits, breads, confectionery and sweets where the weight may be expressed as average net weight or minimum net weight.
- The batch number or lot number or code number may be declared either in numericals or alphabets or in combination, preceded by the words "Batch No." or "Batch" or "Lot No." or "Lot" or any distinguishing prefix.
- The month and year in which the product was manufactured or pre-packed is to be declared except in case of carbonated water containers and packages of biscuits containing 60 grams to 120 grams and packages of food weighing less than 60 grams, bread, milk and for all packages of irradiated food to bear the following declaration and logo:
- The package should declare: The month and year in capital letters up to which the product is best for consumption, in the following manner, namely:

"BEST BEFORE .... MONTHS AND YEAR" or

"BEST BEFORE .... MONTHS FROM PACKAGING" or

"BEST BEFORE .... MONTHS FROM MANUFACTURE" or

"BEST BEFORE UPTO MONTH AND YEAR ...."

PROCESSED BY IRRADIATION METHOD

DATE OF IRRADIATION

In case of package or bottle containing sterilised or ultra high temperature treated milk,soya milk, flavoured milk, any package containing bread, dhokla, bhelpuri, pizza, doughnuts, khoa, paneer, or any uncanned package of fruits, vegetable, meat, fish or any other like commodity, the declaration be made as follows:

" BEST BEFORE .... DATE/MONTH/YEAR" or

" BEST BEFORE ... DAYS FROM PACKAGING".

- The declaration to be made on packages of infant milk substitute and infant milk as per rule 37B states that:

  Every container of infant milk substitute or infant food or any label affixed thereto shall indicate in a clear, conspicuous and in an easily readable manner, the words "IMPORTANT NOTICE" in capital letters and indicating thereunder the following particulars, namely:

  (a) a statement "MOTHER'S MILK IS BEST FOR YOUR BABY" in capital letters. The types of letters used shall not be less than five millimetres and the text of such statement shall be in the Central Penal of every container of infant milk substitute or infant food or any label affixed thereto. The colour of the text printed or used shall be different from that of the background of the label, container or the advertisement, as the case may be. In case of infant food, a statement indicating "infant food shall be introduced only after four months of age" shall also be given.

  (b) a statement that infant milk substitute of infant food should be used only on the advice of a health worker as to the need for its use and the proper method of its use.

  (c) a warning that infant milk substitute or infant food is not the sole source of nourishment of an infant.

  (d) a statement indicating the process of manufacture (spray or roller dried) except in case of infant food, instruction for appropriate and hygienic preparation including cleaning of utensils, bottles and teats and warning against health hazards of inappropriate preparations, as under: "Warning/caution-Careful and hygienic preparation of infant food/infant milk substitute is most essential for health. Do not use fewer scoops than directed since diluted feeding will not provide adequate nutrients needed by your infant. Do not use more scoops than directed since concentrated feed will not provide the water needed by your infant".

  (e) the approximate composition of nutrients per 100 grams of the product including its energy value in Kilo Calories/Joules.

  (f) the storage condition specifically stating "store in a cool and dry place in an air tight container" or the like.

  (g) the feeding chart and directions for use and instruction for discarding left over feed

(h) instruction for use of measuring scoop (level or heaped) and the quantity per scoop (scoop to be given with pack)

(i) indicating the Batch No., Month and Year of its manufacture and month and year before which it is to be consumed

(j) the protein efficiency ratio (PER) which shall be minimum 2.5 if the product other than infant milk substitute is claimed to have higher quality protein

Eforcement of the PFA

Under the PFA, Food Inspectors are appointed by the State governments. They are often a part of the Food and Drug Administration or Local Health Authority. The Food Inspector has the power to take a sample of the food from the place of manufacture, storage or from seller and send it to a Public Analyst for testing. Public Analysts have been created under the Act to analyse sample of article of food sent to them. A Food Inspector who intends to take a sample has to disclose his identity and inform the retailer his intention of taking a particular product as a sample for analysis. The Food Inspector takes three samples, which are to be sealed and labeled. He sends one sample with a memorandum to a Public Analyst. The other two samples are deposited in the office of the department to which the Food Inspector belongs. The Public Analyst sends his report. If the Analyst's report declares that the sample is not in conformity with the provisions of the PFA, the Food Inspector initiates prosecution of the PFA, the Food Inspector initiates prosecution in the court of a first class magistrate. The Food Inspector while taking sample asks the retailer to disclose the name of the wholesaler/distributor. From the package, the Food Inspector also gets to know th name of the manufacturer and distributor. Thus, he knows the entire chain. The Food Inspector can, and often does, make all the parties in the chain accused in the first instance itself. Along with initiating prosecution, the Food Inspector sends letters to manufacturer, distributor and wholesaler. The letter informs that a case has been initiated and that the accused can make an application before the court to have their sample retested by a Central Food Laboratory within ten days from the receipt of the letter. The analysis by the Central Food Laboratory is considered superior to the report of the Public Analyst. It is a right of the accused to get a sample re-tested from a Central Food Laboratory.

If an accused makes an application to the Magistrate, the court directs the department to produce the remaining two samples. After inspecting the seal, the court sends one sample to a Central Food Laboratory. The Central Food Laboratory sends the report. If the report declares the sample to be in conformity with the provisions of the PFA and Rules, the court discharges the case. If the sample fails, trial by the magistrate starts. Corporate bodies like companies, co-operatives or firms are also persons in the eyes of the law. These can be prosecuted and punished as corporate bodies under the Act. A fine can always be paid out of their corporate account. However, a company or co-operative is not a real person, who can be imprisoned. There have to be specific persons who can be held responsible. The PFA makes provisions that corporate bodies can authorise a person, like a director, manager or secretary to exercise all such powers and take all steps to prevent food adulteration and inform the local (health) authority of such an authorisation. The authorized person is called a PFA nominee. The nominee represents the organisation for all matters dealing with the PFA. He is held responsible for any violation committed by the firm. If an organisation has not appointed a nominee, the court holds the person who was responsible to the corporate body for the commission or omission of the action, which led to the violation of the PFA.

The PFA is a legislation of the 1950s. Its dominant horizon is articles of food being sold loose. In the past five decades, the nature of food industry has changed. The market for processed and pre-packed food has expanded tremendously. The food industry deploys sophisticated and expensive food processing technology. There has been a revolution in creation and use of newer packaging materials to give protection to articles of food. The law needs to be revised to take stock of these practices. Law should not only continue to deter food adulterators, but it should also be revised to be optimum in its effect, severe on violators and facilitators to others.

The Fruit Products Order (FPO)

The Fruit Products Order is concerned with fruit and vegetable products including synthetic beverages, syrups, sharbats and vinegar. The objective of this law is mainly to regulate the quality and hygiene of these products. The important labeling rule under FPO is that all labels should have the approval of the authorities concerned, and carry the license number allotted. When a bottle is used as the package, it should be so sealed that it cannot be opened without destroying the license number, and the special identification mark of the manufacturer should be displayed on the top or neck of the bottle. The batch / code number along with the date of manufacturing should also be declared. As contained in PFA, FPO also prohibits use of any statement, design or device, which is false or misleading concerning the fruit product. Synthetic products associated with fruits and vegetables should clearly be marked "SYNTHETIC" and the word, "SYNTHETIC", whenever used, should be as bold and in the same size and colour of the letters used for the name of the product, and should immediately precede such name.

Meat Food Products Order (MFPO)

Meat Food Products Order, similar to FPO, regulates the licensing and labeling of all meat products. All labels have got to be approved by the licensing authority, and the license number and category of manufacturer should be declared on the label. The name of the product, always a common name understood by the consumer, should be given along with net quantity. Trade names should have prior approval of the licensing authority. When any preservative or colouring agent is used, a statement to that effect should be given. When permitted artificial flavouring agent is used, the words, "Artificially Flavoured", should appear on the label in prominent letters and in continuance of the name of the product. The list of ingredients should also be given. Terms which may bear some geographical significance with reference to a locality other than in which either the factory is located, or the product is manufactured, can be given on the label after being qualified by the word, "STYLE", "BRAND", or

"TYPE", as the case may be. No statement, word, picture or design, which may convey a false impression or give a false indication of origin or quality, can appear on the label.

Agricultural Grading and Marking (AGMARK) Rules

Agmark rules relate to the quality specifications and needs of certain agricultural products to be eligible for Agmark Certification. They also specify the type of packages that can be used for various products and labeling declarations that have to be given. Some of the food products that have been covered under these rules are edible nuts, ghee, honey, pulses, spices and condiments and vegetable oil.

The Edible Oil Packaging (Regulation) Order, 1998

In order to ensure availability of safe and quality edible oil in packed form, the Central Government promulgated on 17th September,1998 a Packaging Order under the Essential Commodities Act, 1955 to make packaging of edible oil, sold in retail, compulsory unless specifically exempted by the concerned State Governments.

Uniform methods for testing the quality of edible oil, including the Thin Layer Chromatography (TLC) method for detection of Argemone oil was prescribed and circulated to all State Governments and manufacturers.

The salient features of the Packaging Order are:

- Edible oil including edible mustard oil will be allowed to be sold only in packed form from 15th December,1998.
- Packers will have to register themselves with a registering authority.
- The packer will have to have his own analytical facilities or adequate arrangements for testing the samples of edible oil to the satisfaction of the Government.
- Only oil which conform to the standards of quality as specified in the Prevention of Food Adulteration Act, 1954 and Rules made thereunder will be allowed to be packed.
- Each container or pack will have to show all relevant particulars so that the consumer is not misled, so also the identity of the packer becomes clear.
- Edible oil shall be packed in conformity with the Standards of Weights and Measures (Packaged Commodities) Rules, 1977, and the Prevention of Food Adulteration Act, 1954 and Rules made there under.
- The State Governments will have power to relax any requirements of the packaging order for meeting special circumstances.

The power for implementation of the Order is basically delegated to the State Governments. The Central Government is aware that the production of edible oils is a highly de-centralised industry. A substantial quantity of oil production is in the small-scale or unorganised sector. Further, a sizeable proportion of the population is living below the poverty line. It may be difficult for them to afford the additional cost of packaged oils. It is in view of these situations that the State Governments have been empowered to exempt any edible oil from the provisions of this order in specific circumstances.

Packaging is being recognised as a major industry in all developing countries. This is not surprising as all the products manufactured or processed are packed in some way or the other, so as to safeguard the interests of the consumer and the society. The laws and regulations that apply to these products are very critical. These laws act as a measure of protection and selfsatisfaction for the customers in terms of quality and quantity.

21

Folding Boards Cartons

Folding Cartons are protective containers made from sheets of paperboard that have been cut and creased for forming into derived shape which facilitate the distribution, sale and use of its contents.

The most popular of the various types of rigid packages.

- Very economical in cost of material as well as fabrication and assembly.
- Being collapsible, it takes a minimum of space in shipment and in storage.
- Finest kinds of printing, embossing and hot-foil stamping can be used to make a very attractive package.
- Adds value and sales appeal to the product.
- Versatile *i.e.* great variety of sizes, styles and other special features possible.

The folding carton created the packaging industry as it is known today, beginning in the late 19th century. Basically, a folding carton is made of paperboard, and is cut, folded, laminated and printed for transport to manufacturers. The cartons are shipped flat to a manufacturer, which has its own machinery to fold the carton into its final shape as a container for a product. The classic example of such a carton is a cereal box.

In the 1840s, cartons were made by hand and held together with tacks and string, and used only for expensive items (such as jewelry). Although Charles Henry Foyle is described by some as the "inventor" of the paper carton, mass production of the cartons was invented, partly by accident, at the Robert Gair Company in Brooklyn, New York. Machinery at the end of the press had been set up carelessly by a pressman, and machinery cut through the material. This ruined the press but gave them an idea:

printing and cutting could be done with one machine. Previously, cutting of printed cardboard had been done manually. From the mistake in 1879, Gair developed a process for mass production of boxes. In 1897, the National Biscuit Company (Nabisco) became the first large company to adopt the new cartons, for Uneeda Biscuits. Other manufacturers soon followed.

With inexpensive packaging now even common items could be placed in a showy carton and each carton became its own advertisement. The product was also protected, and the contents had a longer shelf life. This trend was to continue with force, through the 20th century. This could be seen as a contributing factor in the so-called 'throwaway' culture of America. The environmental impact of product packaging has gained attention from consumers and businesses alike, and this awareness has created a steady trend since the mid to late 1990s, on the part of manufacturers, to use recycled material and / or reduce overall materials usage.

Folding cartons are now a $50 billion industry. Typically, cylinder board made from pulp from reprocessed scrap paper is used for most packages. Cartons for food are made from a higher grade and lighter solid sulfate board. Because of the limitations of cutting machinery, the thickness of the board is limited to 0.81 mm (0.032 in), and folding cartons are generally limited to holding a few pounds or kilograms of material.

Folding carton manufacture is a combination of science, engineering and art; if you wanted to build a state-of-the-art factory from scratch, it would require an investment of tens of millions of dollars in complex, sophisticated equipment, not to mention a commitment to finding the highly skilled pressmen and craftsmen with the knowledge necessary to turn an idea into a finished folding carton.

the folding carton is erected either by hand or automatically after which it is filled with product. Freight costs to the end user are therefore much less than rigid packaging, which must be shipped already formed.

What it is Not

When you think of folding cartons, you are not generally thinking of shoe boxes or jewelry boxes. For the most part, this type of packaging, called a rigid, or setup box, is one of the oldest forms of packaging. The panels of these boxes are made from thick, unbending chipboard which are taped or otherwise attached to each other to form the container. They cannot be folded flat, but are constructed in their final, three-dimensional shape. These boxes usually have a top and a bottom, and may be wrapped with printed or foiled paper Folding cartons are also if rarely used as shipping containers – this is the province of corrugated boxes.

What it is Made From

Paperboard is a grade of fiberboard specially manufactured to withstand creasing. In other words, it will bend or fold without cracking. It is often smoothed and coated to give it a superior printing surface. The behemoths that make paperboard can be so long it is difficult to see the entire machine while standing at either end. There are two basic grades of paperboard: virgin, made with at least 80 per cent previously unused wood pulp, and recycled, board made from re-pulped, previously used paperboard.

Almost all packaging made with any grade paperboard is recyclable, making this packaging substrate the most sustainable and environmentally friendly available.

Folding cartons may also be made from narrowly fluted corrugated sheets. Corrugated is made from a fluted medium sandwiched between two liners. To put these in perspective, C flute is the most common type of corrugated used in shipping containers. This material is too thick to produce folding cartons, so production using this substrate is limited to the narrower gauges. It is also possible to combine the strength of corrugated with the graphic superiority of paperboard through the use of single face laminating. The typical brown top liner of corrugated is replaced by a thin sheet of bright white paperboard called solid bleached sulphate. The paperboard is printed, then laminated to the flute, after which the sheet may be converted like any other folding carton.

Stages in the Manufacture of a Folding Carton

- Package Design
- Layout
- Die Making
- Sheeting
- Printing
- Cutting and Creasing
- Waste Stripping and Recycling
- Other Processes
- Gluing

Package Design

There are two decisions that must be made when designing a folding carton to market, ship and protect the product inside: what shape will the box take, and what *graphics* (images other than type) will be added to identify and sell the product? With advances in computer technology and CAD (Computer Aided Design), a client may see a full-color, three-dimensional mock-up of the finished product before any production costs are incurred. The basic folding carton begins as a simple, two-dimensional, four-sided tube with an extension at one end, the glue flap, which allows the tube to be folded into its three- dimensional shape. Various types of closures, such as tuck tops and dust flaps (as seen in Figure on next page) may be added on the top and bottom of the tube to create the final design.

Once the shape of the carton has been determined, the graphics and type which will identify and market the product are registered (aligned) to each panel. After the printed sample is approved by the customer, the design moves on to Layout.

Layout

The paperboard mill slits these very wide rolls into smaller ones that will fit on the converter's equipment ("converter" is another term for the folding carton

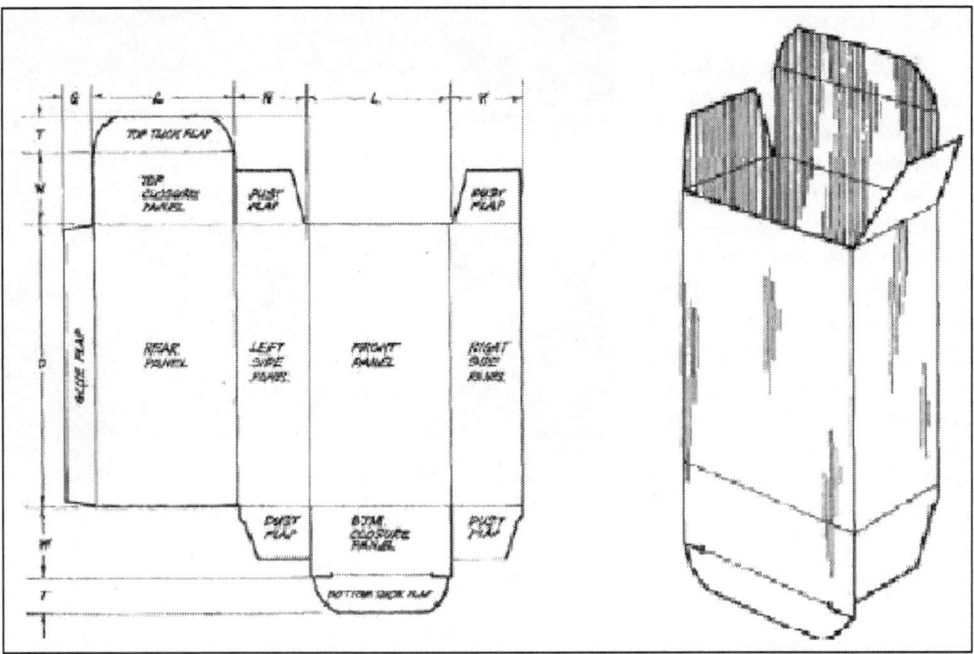

manufacturer). Cartons are not made one at a time, but rather in large sheets or in a continuous stream called a web. Whether printing will be accomplished on a web-fed or sheet-fed press, the individual carton must laid out in a pattern that attains the best use of *board* (an abbreviated way of saying "paperboard") and the least amount of waste. This raw material may easily account for 35 per cent or more of the total cost of the carton, so efficient utilization is critical. Figure below shows a press sheet which will print eight cartons, 4 "up" of two different styles.

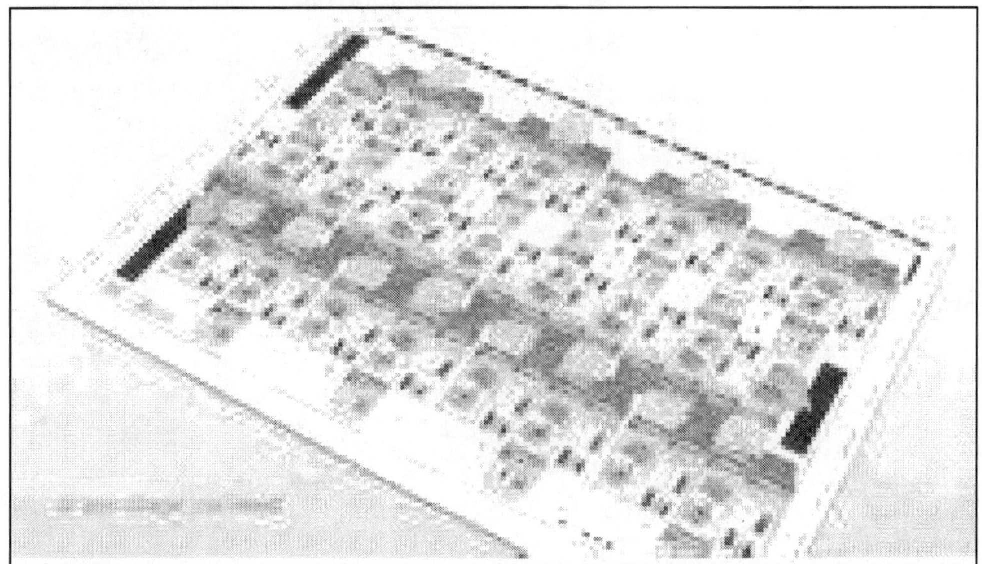

Die Making

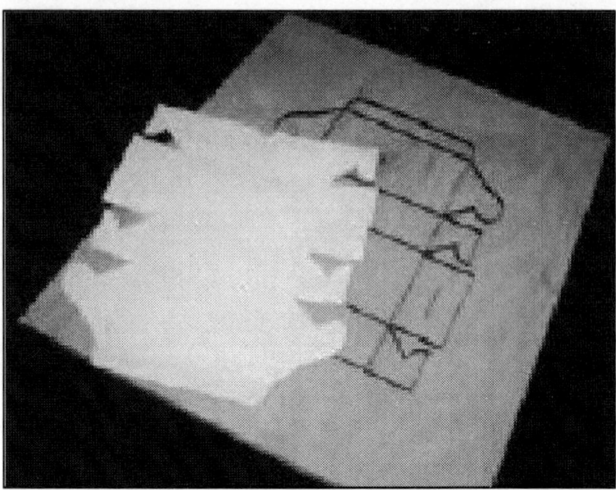

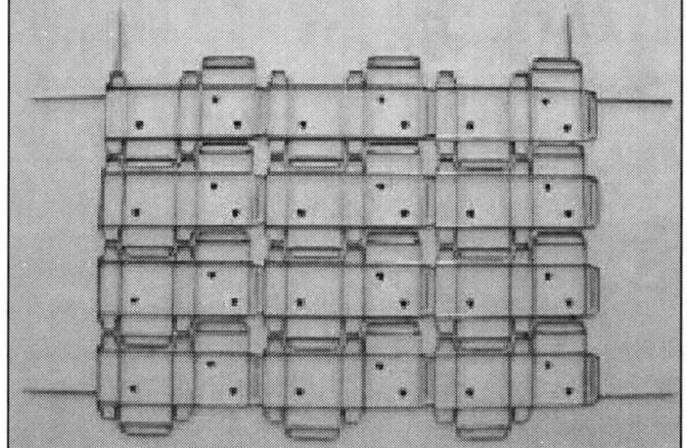

In the modern folding carton plant, the same computer program that generates the structure of the box and the subsequent production layout is used by the die maker to create a steel rule die (Figure). Figure shows a 12-up die. Many die boards today are cut by laser, insuring a much higher degree of accuracy and consistency between dies. Steel rule knives are then bent and cut to the required shape and fit into the slots burned by the laser. Where the design calls for a crease, the steel rule has a rounded edge that creates a score around which the carton will fold. Also, note the "nesting" of cartons in Figure 11. The bottom tucks of the top two rows of cartons are laid into the recesses formed by the boxes below them.

Wherever possible, the boxmaker will maximize sheet utilization, creating as little waste as possible. Rotary dies (3rd Figure right) are used depending on the die cutting equipment to be used.

Sheeting

If paperboard is received into a sheet-fed carton plant in rolls, it must be sheeted in order to be able to be fed into the presses. This is the first step in the converting process, where raw material, paperboard, is converted into packaging.

A Word About "Grain" The slurry which enters the "wet end" of the paper making machine is about 99 per cent water and 1 per cent wood fiber. Through the giant length of the machine, as water is pressed and dried from the wood pulp, the individual solid fibers are aligned in the direction the slurry is traveling as it makes its way toward the "dry end," where the resulting paperboard contains about 5 per cent moisture. This alignment is called *grain,* and in the carton shop, folding a score on a box that runs *with the grain* is much easier than folding it *across the grain*. Back at the sheeter, a mill roll 28" wide is being pulled off an unwind stand and

cut into sheets 28" X 40". The grain direction always runs in the same direction in which the board was made, which in the sheeter is called the *cutoff,* in this case the 40" length of the sheet. For consistency, grain direction is always expressed as the *second* dimension of a sheet of paperboard. A good packaging designer will use grain to his or her advantage to make a box that sets up easily and closes securely. The way a box is laid out on a sheet can mean the difference between a tuck flap closure that locks securely or one that won't lay flat and wants to pop open when it shouldn't.

Printing

Screen Printing Basic

Traditional printing presses cannot print continuous tones: smooth graduations from light to dark tones of the same color. The way printers have gotten around this problem is to create an optical illusion called a ***halftone***. Look at any photograph in a newspaper through a magnifying glass and you will see that it is composed of thousands of dots. The halftone converts a continuous tone into a pattern of solid dots. The dots are either very small for highlighted areas or very large (relatively speaking) for shadows. (In stochastic screens, the dots are all the same size, but randomly distributed, producing a similar effect.) Viewed at a distance, the eye cannot distinguish between these tiny printed motes, and fills in the missing information to "reform" the image as a continuous tone

Color images are created in the same way. A color photograph of continuous tone is separated and screened into the four ***process*** colors: cyan, magenta, yellow and black (CMYK). By combining various combinations of CMYK dots of differing sizes, many more colors may be simulated. These are called ***manufactured*** colors. However, some colors, such as grays, reds and violets do not manufacture perfectly from a combination of the four CMYK inks. In this case, these colors may be matched by a single ***spot color*** ink, providing there are enough units on the press. The number of halftones that are printed in a linear inch is called ***line screen***. It's value is expressed as Lines per Inch (LPI). Newspapers, which are printed on porous, lower quality paper may require as little as a 70 line screen to reproduce a reasonably sharp image, but high quality images printed on packaging will typically be printed between 133 and 150 LPI. Glossy, color magazine print may go as high as 300 LPI or more.

Pre-press

How does original artwork wind up on the finished folding carton? In the old days (the distant 1980's), just about every printer using offset lithography, which will be described below, made films from four-color negatives which were then exposed onto printing plates, and they in turn were *hung* (attached to the plate cylinder) on the press, one printing unit for each color. If the customer required a color that could not be manufactured from the basic four (cyan, magenta, yellow, black) and the printer only had a printing press with four stations, this *spot* color would require a second pass once the first four colors were laid down. In the modern pre-press room, the need for the intermediary step of creating films israpidly disappearing. Computer-to-Plate (CtP) technology allows the image (now computerized with desktop publishing software) to be output directly to the printing plate.The next step, in limited use as of this writing, is Digital Imaging (DI), which removes films and plates altogether. A computer attached to the press applies the multi-color image to the paperboard in much the same way that toner is applied to create an image in an office copy machine.

It is printed in any one of the Four Primary Printing Processes. The most common is offset and flexography.

Cutting and Creasing

Once a sheet or web of paperboard has been printed, the shape of the container created in the design phase is used to manufacture the cutting die (either flatbed or rotary). Back in the early days of folding carton manufacture, printing was accomplished primarily on flatbed presses. The sheet was hand-fed into the press, wrapped around a rotating cylinder and printed as it passed over a reciprocating letterpress die moving back and forth beneath it.

By removing the ink rollers and replacing the printing die with a cutting and creasing die, this same configuration could be used to cut and score the dried, printed sheet. Today, in the modern reciprocal diecutter, the die moves up and down, not back and forth. The pile of printed sheets is fed from the right, registered on the feed board (the blue arrow) and pulled into the cutting station. The steel rule die is secured to a reciprocating platen which descends with precise timing and many hundred tons of pressure onto the sheet. This presses the sheet onto a counter plate which does two things: provides a stable surface against which the knives in the die can cut, and holds the female channels that receive the male scores in the die. These channels help create scores that will not crack upon folding when the carton is glued and subsequently erected and filled with product. Embossing or debossing may also be accomplished in the cutting and creasing station After the sheet leaves this section of the press, it is held together by tiny nicks in the die that keep it from falling apart until it reaches the stripping station. At this point, all internal and side waste is trimmed away. Some presses may even deliver separated and counted individual piles of cartons, neatly stacked and ready for the next converting process. A press this size, which can handle a sheet up to 44 X 64 inches, may reach speeds as high as 7,000 sheets per hour.

Stripping of Waste

Whether the folding carton is printed on a web or sheet-fed press, internal and external waste must be stripped away before additional finishing is completed. In some cases, this is done after diecutting is complete. Pallet loads of sheets held together by nicks in the die are moved to an area where waste is manually removed either with hand-held hammers and chisels, or even jack hammers. More and more frequently, internal carton waste and trim from the edges of the sheet are automatically removed as part of the cutting and creasing process.

Other Finishing

Several other visual and practical enhancements may be made to folding cartons before they are glued, packed and shipped to the customer. These may include, among others:

- Foil Stamping
- Embossing / Debossing
- Windowing

Foil Stamping

The consumer's eye is easily caught by packaging that stands out on the shelf. One way to accomplish this is by applying bright foil to all or part of the box. More than in just gold or silver, foil now comes in a dizzying array of colors and patterns, even lasered or holographic,

like the boxes on the right. This process is also known as hot stamping, since heat applied to a metal plate with an engraved image improves the release of the foil from its carrier roll as it is pressed onto the paperboard. Note that the image in this brass die is reversed; it will be right-reading once the foil is transferred onto the substrate. Foil stamping presses may be either simple, hand-fed machines or sophisticated, high speed production centers

Embossing/Debossing

Embossing places a raised image on a box. Debossing presses an the image into the surface of the paperboard. A score that will allow a carton to fold without cracking requires a female channel into which the steel creasing rule will press the paperboard. In the same way, the more elaborate images pressed into a box panel for debossing or embossing requires both a raised male die and a recessed female die.

Debossing dies may be imbedded in the folding carton die itself so that when the platen descends, the image is pressed into the paperboard. If the embossing die is placed on the counter plate, the printed side of the sheet will be pushed up, forming a raised, or embossed image.

Windowing

Sometimes there's no substitute for seeing what's inside the box. But placing and gluing a rectangular patch of window film in precisely the right spot over a cutout on a folding carton is no simple matter.

Finishing

Up until this point, the folding carton has remained a flat, essentially two dimensional blank, unusable by the customer. The final box must be folded, glued and shipped flat. When it reaches the customer it will be opened, formed into a tube and filled with product. In order for this to happen, adhesive must be applied to one or more "manufacturer's seals." Gluing can be either timed or untimed. In untimed gluing, adhesive need only be applied to the strip lying outside the interior of the box. When this is the case, cartons can travel as fast as the belt carrying them through the gluer will allow. The glue flaps of these cartons pass over a rotating wheel immersed in a bath of adhesive after which the box is folded upon itself and sealed.

However, as application systems and gluer technology has become more advanced, complex patterns of adhesive can be applied without sacrificing productivity. Cold or hot adhesive may be applied by ejecting precise beads of glue wherever they are required machine, scores are pre-broken, adhesive is applied and glued panels are folded together.

Before the boxes can be packed in corrugated shippers at the end of the line, however, they must pass through a compression section which not only gives the adhesive time to set up, but reduces "fluff," the tendency for the box to spring open, allowing for maximum pack counts when shipping.

Summary

The design and manufacture of a folding carton is a combination of both art and technology. The investment in talent and equipment is substantial, but the result is a package that protects and markets consumer products like no other.

Extrusion Coating

Extrusion coating is a process of coating thermoplastic material onto a substrate such as woven fabric, paper, paperboard aluminum foil, PET, BOPP film etc. The resins most commonly used are polyolefins such as polyethylene, ionomers, ethylene vinyl acetate copolymers and polypropylene.

Advatanges of extrusion coating are:

As a Process

- Double sided coating can be done to achieve desired properties.
- Solvent/adhesive free
- Thickness of the coating can be varied depending on the end use
- Higher line speeds

As an End Product

- Provides moisture barrier properties
- Avoid direct contact of content to the substrate
- Provide heat sealable characteristics
- Reduces loss of content

Process of Extrusion Coating

Extrusion coating is a process in which a substrate is coated with an accurately metered film of molten plastics. Extruded thin molten film is pulled down on a substrate and into the nip between the chill roll and pressure roll below the die (see figure) The pressure between these two rolls forces the plastic onto the surface moving at a speed faster than the extruded film and drawing the film to the required thickness.

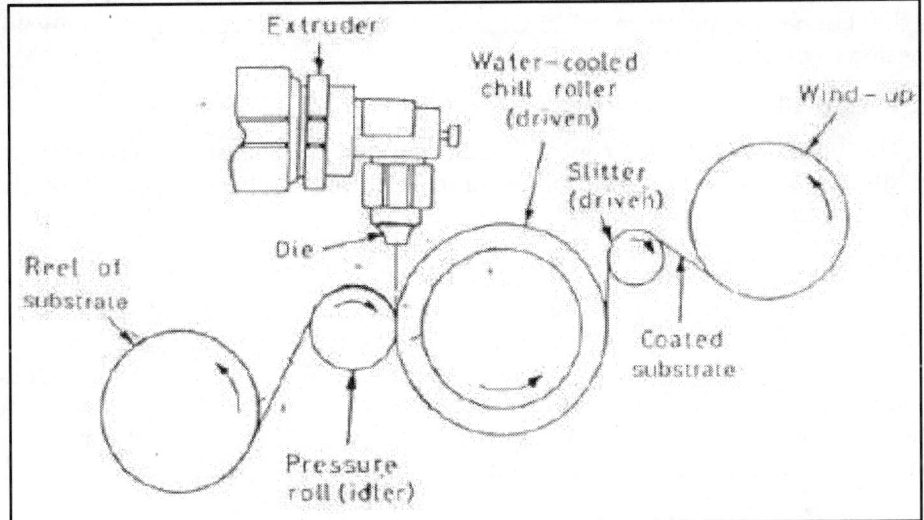

Line Diagram of Extrusion Coating Process

Extrusion coating line consists of:

1) Extruder
2) Die
3) Chrome plated chilled roll
4) Back up roll
5) Winders/Unwinders

An extruder converts solid pellets of thermoplastic resin into a uniform, homogeneous melt. Extruders with L/D ratio of 24: 1 to 26: 1 are normally used for LDPE whereas for LLDPE, L/D ratio of 26: 1 or above is recommended for better results. Die commonly used are flat, coathanger type with adjustable die lip. Melt extruded through the die comes in the nip formed by chrome plated chilled roll and soft rubber coated back roll.

A large diameter chilled roll provides enough contact time to cool the product sufficiently to allow the film to be stripped from the roll without damage. This cooling requirement is often the limiting factor in the running speed for a particular product. The surface of the chilled roll may be bright, dull, matte, embossed depending upon the desired surface characteristics.

Factors Influencing Coating Performance

1. Melt Temp.

Extrusion coating lines are normally operated at higher temperature. The resins for extrusion coating contains lower level of antioxidants. Higher processing temperatures oxidizes the polymer partially and promotes good bonding to the substrate. Further it lowers melt viscosity promoting uniform flow of melt through the die. Higher temperature (>300°C) is necessary to achieve acceptable level of adhesion to the substrate at high line speeds.

2. Die Gap

The die gap is the critical control point on any extrusion coating line. The slit die with die opening of 0.5 mm to 0.8 mm is commonly used. Higher die gaps give higher flow variation.

3. Air Gap

The gap between the die exit and substrate on chilled roll is called air gap which need to be optimized with respect to coating material. Some oxidation of the melt takes place after exit from the die, facilitating adhesion of polymer to the substrate. Higher gap leads to higher necking and cooling of the melt, leading to improper adhesion of polymer to the substrate. For LLDPE/LDPE (1) blend, the air gap is normally maintained between 30 - 40 mm.

4. Chilled Roll Temperature

It has influence on coating adhesion and stiffness. Optimum chilled roll temperature for polyethylene is 30°C.

Material Selection for Extrusion Coating

The resin most commonly used for extrusion coating are LDPE, LLDPE and their binary blends. Each polymer offers unique advantages over the other and their binary blend strikes the balance of properties. Due to presence of long chain branching and broader molecular weight distribution in LDPE, it shows a good processability, superior hot melt strength. Further, LDPE shows a phenomenon called 'strain hardening' by virtue of which elongation visciosity increases with the line speed. Hence draw down to thinner gauges are possible at higher line speed without appreciable neck-in.

Extrusion coating grades with LLDPE have shown that neck-in is high and increases with line speed. The increase in neck-in is attributed to the drop in elongational viscosity with increase in line speed. This phenomenon is termed as 'tension thinning and is due to narrow molecular weight distribution and absence of long chain branching.

However LLDPE held a new distinct advantages over LDPE extrusion coating grade. The main advantage of LLDPE over LDPE is high mechanical strength which allows down gauging. The relatively lower thermo-oxidative stability of LLDPE over LDPE leads former to oxidize faster resulting in better adhesion to non porous substrates like aluminium foil and tapes. A marked improvement in sealing properties is observed when LLDPE is blended with LDPE.

The most practical method to utilize LLDPE in extrusion coating is by blending with highly branched LDPE> Blending of polyolefins generally has an additive effect. LLDPE rich blends offer much better mechanical properties and sealing properties as compared to those of LDPE. The processing characteristics of such blends can be enhanced further by using narrower die gap (0.4 mm - 0.6. mm)

Resin Characteristics for Extrusion Coating

- Higher melt index-required for better adhesion
- Additive free
- Appropriate amount of antioxidants
- Reduce neck-in properties
- Good adhesion to substrate at higher line speeds

Coding

Coding is a critical activity in the packaging process. In case of pharmaceuticals it is of utmost relevance. Coding is the statutory regulatory requirement. Pre pack goods for consumer consumption need to be coded indicating details like manufacturing date, expiry date, manufacturing license and in case of India maximum retail price (all inclusive). It is possible to print it on the label along with the other details.

However it does not make a business sense as a product manufacturing is always done batch wise and the manufacturing date varies. In order to overcome this these statutory details are printed on to the label or pack either just before filling or after filling or sealing the pack. In fact, on the pack label there are specific areas located for printing this data. In India the statutory regulations clearly specify a contrast between the back ground and the printing data. This is why in the packs a white patch is provided wherein this data gets printed.

This process is generally on line in any packaging unit.

In general coding can be broadly classified in to two categories

1. Contact coder
2. Non contact coder

Contact Coding

In the earlier days this used to be done using rubber stereos manually this is similar to rubber stamping. Later as production lines were automated this activity was also automated. Using rubber stereos on the automated lines.

The disadvantages using contact coder are:

1. Smudging of the letters due to uneven pressure in manual coding
2. Smudging of letters due to ware and tare in automated lines

3. Smudging of letters due to slow drying
4. Illegible printing due to drying of the ink
5. Repeated change of the stereos whenever the batch or date changes
6. Development of suitable drying ink for each line depending on its speed
7. Possible errors due to improper fitting of the letters
8. Possible errors due to falling of the letters on the automated line

All these defects can be identified only when the printing is done. This leads to rejection of the final packed product due to improper printing which gain is a major loss.

Hence the concept of non contact coding was developed. In this case the print head does not come in contact with the substrate on which printing is done. This is inception of inkjet coders or inkjet printers.

Origins

Although inkjet printers were first mass-produced in the 1980s, it was only in the 1990s that prices dropped low enough for that technology to be brought into the mass consumer market. Canon claims to have invented what it calls 'bubble jet' technology in 1977, when a researcher accidentally touched an ink-filled syringe with a hot soldering iron and the heat forced a drop of ink out of the needle. And so began the development of a new printing method.

Inkjet printers have made rapid technological advances in recent years. First, the three-color printer succeeded in making color inkjet printing an affordable option; but as the superior four-color models became cheaper to produce and sell, it wound up being the standard and users' choice.

Inkjet printing has two chief benefits over laser printers: lower printer cost and color-printing capabilities. But while inkjet printers are priced much less than laser printers, they are actually more expensive to use and maintain. Cartridges need to be changed more frequently and the special coated paper required to produce high-quality output is very expensive. At a cost per page level, inkjet printing costs about 10 times more than laser printing.

Inkjet printers technology development starts in the early 1960s. The first inkjet printing device was patented by Siemens in 1951, which led to the introduction of one of the first inkjet chart recorders. The continuous inkjet printer technology was developed later by IBM in the 1970s. The continuous inkjet technology basis is to deflect and control a continuous inkjet droplet stream direction onto the printed media or into a gutter for recirculation by applying an electric field to previously charged inkjet droplets

The drop-on-demand inkjet printer technology was led to the market in 1977 when Siemens introduced the PT-80 serial character printer. The drop-on-demand printer ejects ink droplets only when they are needed to print on the media. This method eliminates the complexity of the hardware required for the continuous inkjet printing technology. In these first inkjet printers ink drops are ejected by a pressure wave created by thc mechanical motion of the piezoelectric ceramic.

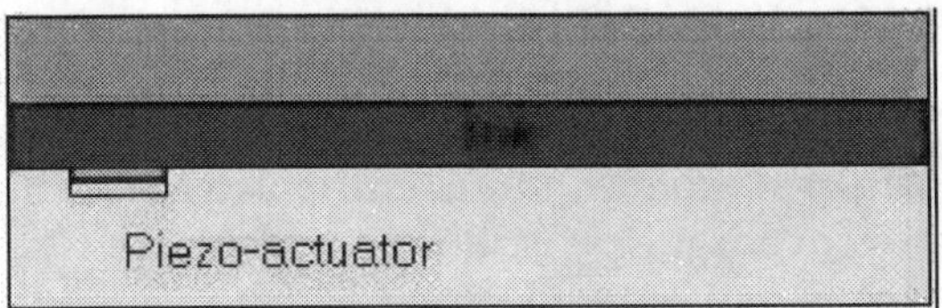

At the same time Canon developed the bubble jet printer technology, a drop-on-demand inkjet printing method where ink drops were ejected from the nozzle by the fast growth of an ink vapor bubble on the top surface of a small heater. Shortly thereafter, Hewlett-Packard independently developed a similar inkjet printing technology and named it thermal inkjet.

The most popular inkjet and bubble-jet printers use serial printing process. Similarly to dot matrix printers, serial inkjet printers use print heads with a number of nozzles arranged in vertical columns. The printing process is the same as in dot matrix printers.

The greatest advantages of inkjet printers are, quiet operation, capability to produce color images even with photographic quality and the low printer prices. The down side is that although inkjet printers are generally cheaper to buy than lasers, they are far more expensive to maintain. When it comes to comparing the cost per page, ink jet printers work out many times more expensive than laser printers. There are some exceptions of course for some heavy-duty industrial printers. From Tallyclaim the T3016 SprintJet prints at only 1/3 of a cent per page.

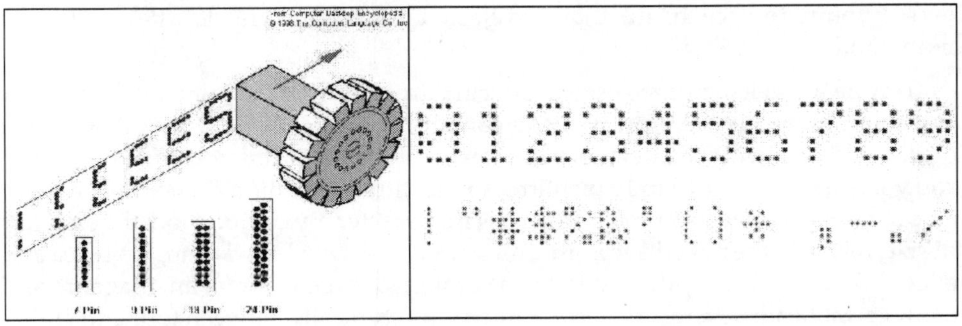

Operation

Inkjet printing, like laser printing, is a non-impact process. Ink is emitted from nozzles while they pass over media. The operation of an inkjet printer is easy to visualize: liquid ink in various colors being squirted onto paper and other media, like plastic film and canvas, to build an image. A print head scans the page in horizontal strips, using the printer's motor assembly to move it from left to right and back again, while the paper is rolled up in vertical steps, again by the printer. A strip (or row) of the image is printed, then the paper moves on, ready for the next strip. To speed things up, the print head doesn't print just a single row of pixels in each pass, but a vertical row of pixels at a time.

For most inkjet printers, the print head takes about half a second to print the strip across a page. On a typical 8 1/2"-wide page, the print head operating at 300 dpi deposits at least 2,475 dots across the page. This translates into an average response time of about 1/5000th of a second. Quite a technological feat! In the future, however, advances will allow for larger print heads with more nozzles firing at faster frequencies, delivering native resolutions of up to 1200dpi and print speeds approaching those of current color laser printers (3 to 4 pages per minute in color, 12 to 14ppm in monochrome). In other words, declining costs for improving technology.

There are several types of inkjet printing. The most common is "drop on demand" (DOD), which means squirting small droplets of ink onto paper through tiny nozzles; like turning a water hose on and off 5,000 times a second. The amount of ink propelled onto the page is determined by the print driver software that dictates which nozzles shoot droplets, and when. The nozzles used in inkjet printers are hairbreadth fine and on early models they became easily clogged. On modern inkjet printers this is rarely a problem, but changing cartridges can still be messy on some machines. Another problem with inkjet technology is a tendency for the ink to smudge immediately after printing, but this, too, has improved drastically during the past few years with the development of new ink compositions.

Thermal Technology

Most inkjets use thermal technology, whereby heat is used to fire ink onto the paper. There are three main stages in this process. The squirt is initiated by heating the ink to create a bubble until the pressure forces it to burst and hit the paper. The bubble then collapses as the element cools, and the resulting vacuum draws ink from the reservoir to replace the ink that was ejected. Canon and Hewlett-Packard favor this method.

Tiny heating elements are used to eject ink droplets from the print head's nozzles. Most thermal inkjets have print heads containing a total of between 300 and 600 nozzles, each about the diameter of a human hair (approx. 70 microns). These deliver drop volumes of around 8 to 10 picolitres (a picolitre is a million millionth of a liter), and dot sizes of between 50 and 60 microns in diameter. By comparison, the smallest dot size visible to the naked eye is around 30 microns. Dye-based cyan, magenta and yellow inks are normally delivered via a combined three-color (cyan, magenta and yellow) print head. Several small color ink drops - typically between four and eight -

are typically combined to deliver a variable dot size. Black ink, which is generally based on bigger pigment molecules, is delivered from a separate print head in larger drop volumes of around 35pl.

Nozzle density, corresponding to the printer's native resolution, varies between 300 and 600 dpi, while enhanced resolutions of 1200 dpi are increasingly becoming available. Print speed is chiefly a function of the frequency with which the nozzles can be made to fire ink drops and the width of the swath printed by the print head. This is usually around 12MHz and half an inch respectively, giving print speeds of between 4 to 8 ppm for monochrome text and 2 to 4 ppm for color text and graphics.

Thermal technology, meanwhile, imposes the limitation that whatever type of ink is used; it must be heat-resistant because the firing process is heat-based. Using heat in thermal printers conversely also creates a need for a cooling, which adds a to the overall length of printing time.

Piezo-Electric Technology

Epson's proprietary inkjet technology uses a Piezo crystal at the rear of the ink reservoir. This is rather like a loudspeaker cone that flexes when an electric current flows through it. So whenever a dot is required, a current is applied to the Piezo element which then flexes and, in so doing, forces a drop of ink out of the nozzle.

There are several advantages to the Piezo method. The process permits greater control over the shape and size of the released ink droplet. The minuscule fluctuations in the crystal allow for smaller droplet sizes and hence higher nozzle density. Unlike thermal technology, the ink does not have to be heated and cooled between each

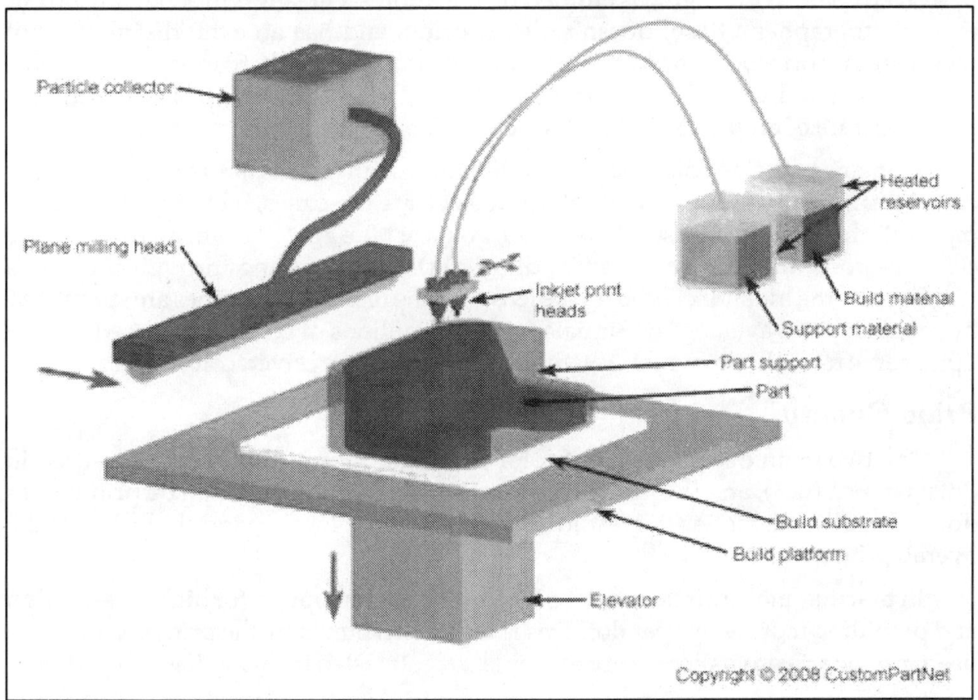

cycle. This saves time, and the ink itself is geared more for its absorption properties than its ability to withstand high temperatures. This enables greater freedom in developing new types of inks.

Epson's newest mainstream inkjets have black print heads with 128 nozzles and color print heads with 192 nozzles (64 for each color), addressing a native resolution of 720 by 720dpi. Because the Piezo process can deliver small and perfectly formed dots with extreme accuracy, Epson is able to offer an enhanced resolution of 1440 by 720dpi (although this is achieved by the print-head making two passes, with a consequent reduction in print speed). The inks that Epson has developed for use with its Piezo technology are solvent-based and extremely quick drying. They penetrate the paper and maintain their shape rather than spreading out on the surface and causing dots to interact with one another. The result is extremely fine print quality, especially on coated or glossy paper.

Color Perception

Visible light falls between 380 Nm(violet) and 780 Nm (red) on the electromagnetic spectrum. White light comprises roughly equal proportions of all the visible wavelengths, and when this light is shined on or through an object, some wavelengths are absorbed while others are reflected or transmitted. It's the reflected or transmitted light that gives the object its perceived color. Leaves, for example, are usually seen as green because chlorophyll absorbs light at the blue and red ends of the spectrum and reflects back the green part in the middle.

The temperature of a light source, measured in Kelvin (K), affects an object's perceived color. White light, as emitted by the fluorescent lamps in a viewing box or by a photographer's flash, doesn't distort colors and has an even distribution of wavelengths corresponding to a temperature of around 6,000 K. Standard light bulbs, on the other hand, emit less light from the blue end of the spectrum, corresponding to a temperature of around 3,000K, which causes objects to appear more yellow.

Humans perceive color via a layer of light-sensitive cells at the back of the eye called the retina. The key retinal cells in the eye are the cones, which contain photo pigments that render them sensitive to red, green or blue light (the other light-sensitive cells, the rods are only activated in dim light). Light passing through the eye is regulated by the iris and focused by the lens onto the retina, where cones are stimulated by the relevant wavelengths. Signals from the millions of cones are passed via the optic nerve to the brain, which assembles them into a perceived color image.

Print Quality

The two main determinants of color print quality are resolution, measured in dots per inch (dpi), and the number of levels or graduations that can be printed per dot. Generally the higher the resolution and the more levels per dot, the better the overall print quality.

In practice, most printers make a trade-off between opting for higher resolution and providing more levels per dot. This is often determined by the printer's intended use. Graphic arts professionals, for example, are interested in maximizing the number of levels per dot to deliver higher 'photographic' image quality, while general business users will require reasonably high resolution so as to achieve good text quality and reasonable image quality.

The simplest type of color printer is a binary device in which the cyan, magenta, yellow and black dots are either "on" (printed) or "off" (not printed), with no intermediate levels possible. If ink dots can be mixed together to make intermediate colors, then a binary CMYK printer can print only eight 'solid' colors (cyan, magenta, yellow, red, green and blue, plus black and white). Clearly this isn't a sufficient palette to deliver quality color printing, which is why there are half tones.

Half toning algorithms divide a printer's native dot resolution into a grid of halftone cells and then turn on varying numbers of dots within these cells in order to mimic a variable dot size. By carefully combining cells containing different proportions of CMYK dots, a half toning printer can 'fool' the human eye into seeing a palette of millions of colors rather than just a few colors.

In continuous tone printing, there's an unlimited palette of solid colors. In practice, 'unlimited' means 16.7 million colors, which is more than the human eye can distinguish. To achieve this, the printer must be able to create and overlay 256 shades per dot per color, which obviously requires precise control over dot creation and placement. Continuous tone printing is largely the province of dye sublimation printers. However, all of the mainstream printing technologies can produce multiple shades (usually between 4 and 16) per dot, allowing them to deliver a richer palette of solid colors and smoother halftones. Such devices are referred to as "contone" printers.

Six-color inkjet printers have now appeared on the market, specifically geared for delivering photographic-quality output. These devices add two further inks - light cyan and light magenta - to make up for current inkjet technology's inability to create very tiny (and therefore light) dots. Six-color inkjets produce more subtle flesh tones and finer color graduations than standard CMYK devices, but are likely to become unnecessary in the future when ink drop volumes are expected to shrink to around 2 to 4 picoletres. Smaller drop sizes will also reduce the amount of half toning required, as a wider range of tiny drops can be combined to create a broader palette of solid colors.

Market-leader Hewlett Packard has consistently advocated the advantages of improving color print quality by increasing the number of colors that can be printed on an individual dot rather than simply increasing dpi, arguing that the latter approach sacrifices both speed and causes problems arising from excess ink - especially on plain paper. In 1996 HP manufactured the first inkjet printer to print more than eight colors (or two drops of ink) on a dot. Its DeskJet 850C being capable of printing up to four drops of ink on a dot. Over the years it has progressively refined its PhotoREt color layering technology to the point where, by late 1999, it was capable of producing an extremely small 5 pl drop size and up to 29 ink drops per dot representing over 3,500 printable colors per dot.

22
Packaging Process

Thermoforming

Thermoforming is a manufacturing process where a plastic sheet is heated to a pliable forming temperature, formed to a specific shape in a mold, and trimmed to create a usable product. The sheet, or "film" when referring to thinner gauges and certain material types, is heated in an oven to a high-enough temperature that it can be stretched into or onto a mold and cooled to a finished shape.

In its simplest form, a small tabletop or lab size machine can be used to heat small cut sections of plastic sheet and stretch it over a mold using vacuum. This method is often used for sample and prototype parts. In complex and high-volume applications, very large production machines are utilized to heat and form the plastic sheet and trim the formed parts from the sheet in a continuous high-speed process, and can produce many thousands of finished parts per hour depending on the machine and mold size and the size of the parts being formed.

Thermoforming differs from injection molding, blow molding, rotational molding, and other forms of processing plastics. Thin-gauge thermoforming is primarily the manufacture of disposable cups, containers, lids, trays, blisters, clamshells, and other products for the food, medical, and general retail industries. Thick-gauge thermoforming includes parts as diverse as vehicle door and dash panels, refrigerator liners, utility vehicle beds, and plastic pallets.

In the most common method of high-volume, continuous thermoforming of thin-gauge products, plastic sheet is fed from a roll or from an extruder into a set of indexing chains that incorporate pins, or spikes, that pierce the sheet and transport it through an oven for heating to forming temperature. The heated sheet then indexes into a form station where a mating mold and pressure-box close on the sheet, with vacuum then applied to remove trapped air and to pull the material into or onto the mold along with pressurized air to form the plastic to the detailed shape of the mold. (Plug-assists are typically used in addition to vacuum in the case of taller, deeper-draw formed parts in order to provide the needed material distribution and thicknesses in the finished parts.) After a short form cycle, a burst of reverse air pressure is actuated from the vacuum side of the mold as the form tooling opens, commonly referred to as air-eject, to break the vacuum and assist the formed parts off of, or out of, the mold. A stripper plate may also be utilized on the mold as it opens for ejection of more detailed parts or those with negative-draft, undercut areas. The sheet containing the formed parts then indexes into a trim station on the same machine, where a die cuts the parts from the remaining sheet web, or indexes into a separate trim press where the formed parts are trimmed.

The sheet web remaining after the formed parts are trimmed is typically wound onto a take-up reel or fed into an inline granulator for recycling.

Most thermoforming companies recycle their scrap and waste plastic, either by compressing in a baling machine or by feeding into a granulator (grinder) and producing ground flake, for sale to reprocessing companies or re-use in their own facility. Frequently, scrap and waste plastic from the thermoforming process is converted back into extruded sheet for forming again.

Thing Gauge and Heavy Gauge Thermoforming

There are two general thermoforming process categories. Sheet thickness less than 1.5 mm (0.060 inches) is usually delivered to the thermoforming machine from rolls or from a sheet extruder. Thin-gauge roll-fed or inline extruded thermoforming applications are dominated by rigid or semi-rigid disposable packaging. Sheet thicknesses greater than 3 mm (0.120 inches) is usually delivered to the forming machine by hand or an auto-feed method already cut to final dimensions. Heavy, or thick-gauge, cut sheet thermoforming applications are primarily used as permanent structural components. There is a small but growing medium gauge market that forms sheet 1.5 mm to 3 mm in thickness.

Heavy-gauge forming utilizes the same basic process as continuous thin-gauge sheet forming, typically draping the heated plastic sheet over a mold. Many heavy-gauge forming applications use vacuum only in the form process, although some use two halves of mating form tooling and include air pressure to help form. Aircraft windscreens and machine gun turret windows spurred the advance of heavy-gauge forming technology during WWII. Heavy gauge parts are used as cosmetic surfaces on permanent structures such as kiosks, automobiles, trucks, medical equipment, material handling equipment, refrigerators, spas, and shower enclosures, and electrical and electronic equipment. Unlike most thin-gauge thermoformed parts, heavy-gauge parts are often hand-worked after forming for trimming to final shape or for additional drilling, cutting, or finishing, depending on the product. Heavy-gauge products typically are of a "permanent" end use nature, while thin-gauge parts are more often designed to be disposable or recyclable and are primarily used to package or contain a food item or product. Heavy gauge thermoforming is typically used for production quantities of 250 to 3000 annually, with lower tooling costs and faster product development than competing plastic technologies like injection molding.

Technology

Thermoforming has benefited from applications of engineering technology although the basic forming process is very similar to what was invented many years ago. Microprocessor and computer controls on more modern machinery allow for greatly increased process control and repeatability of same-job setups from one production run to the next, usually with the ability to save oven heater and process timing settings between jobs. The ability to place formed sheet into an inline trim station for more precise trim registration has been hugely improved due to the common use of electric servo motors for chain indexing versus air cylinders, gear racks, and clutches on older machines. Electric servo motors are also used on some modern and more sophisticated forming machines for actuation of the machine platens where form and trim tooling are mounted, rather than air cylinders which have traditionally been the industry standard, giving more precise control over closing and opening speeds and timing of the tooling. Quartz and radiant-panel oven heaters generally provide more precise and thorough sheet heating over older cal-rod type heaters, and better allow for zoning of ovens into areas of adjustable heat.

An integral part of the thermoforming process is the tooling which is specific to each part that is to be produced. Thin gauge thermoforming as described above is almost always performed on in-line machines and typically requires molds, plug assists, pressure boxes and all mounting plates as well as the trim tooling and stacker parts that pertain to the job. Thick or heavy gauge thermoforming also requires tooling specific to each part, but because the part size can be very large, the molds can be cast aluminum or some other composite material as well as machined aluminum as in thin gauge. Typically thick gauge parts must be trimmed on CNC routers or hand trimmed using saws or hand routers. Even the most sophisticated thermoforming machine is limited to the quality of the tooling. Some large thermoforming manufacturers choose to have design and tool making facilities in house while others will rely on outside tool-making shops to build the tooling.

Types of Moulds

Wood Patterns

Wood patterns are generally the first stage to a thermoforming project. They are relatively inexpensive and allow the customer to makes changes to their design very easily. The number of samples that one is able to get from a wood pattern depends on the size of the part and the thickness of the material. Typically, wood patterns are used to gauge general functionality of both the part and the thickness of the material. Once the specifications of the part have been met, the wood pattern is then used to create a ceramic composite mold, or cast aluminum mold for regular production.

Cast Aluminum Molds

Cast aluminum molds are cast at a foundry and typically have temperature control lines running through them. This helps to regulate the heat of the plastic being formed as well as speed up the production process. Aluminum molds can be male or female in nature and can also be used in pressure forming applications. The main drawback with this type of mold is cost.

Machined Aluminum Molds

Machined aluminum molds are like cast aluminum except they are cut out of a solid block of aluminum using a CNC machine and some sort of CAD program. Typically machined aluminum is used for shallow draw parts out of thin gauge material. Applications may include packaging as well as trays. Again, cost is a significant factor with this type of tooling.

Composite Molds

Composite molds are a lower cost alternative to cast or machined aluminum molds. Composite molds are typically made from filled resins that start as a liquid and harden with time. Depending on the application, composite molds last a relatively long time producing high quality parts. Within the category of composite molds, the subset of "Ceramic" molds has consistently proven to be the most durable. While not temperature controlled, these molds can run nearly as fast as Cast or Machined aluminum, yet at a substantially lower price point. Suitable for all but the highest volume production and strictest tolerances

Thermoforming is a generic term for the manufacturing of plastic components through the vacuum and/or pressure forming processes. A simplistic overview of the single-sheet thermoforming process consists of heating extruded plastic sheet and forming the sheet over a male mold or into a female mold. Depending on what type of mold a customer selects, the thermoforming process allows the customer the ability to receive a part with the same aesthetic properties as an injection-molded part at a fraction of the tooling expense involved in injection molding.

Some of these same principles apply to another thermoforming processes: twin sheeting. Twin-sheeting is heating two sheets of plastic and then forming and sealing the two sheets of plastic together during the forming process. In some cases, the twin-sheeting process allows the customer to receive 1) a higher quality, more aesthetic part, 2) a two color part, if required, and 3) a part with an assembly device trapped inside as the part is being formed.

Vacuum Forming

The most common method with the simplest mold. The sheets adhere to the mold using atmospheric pressure. Various other versions of the process are available, using pre-blowing (balloon), negative forming (plug assisted), reverse blow molding, etc.

Pressure Forming

A blowing bell is combined with mold, in order to increase, through the use of compressed air, the adherence of the sheet onto mold. Indispensable for tenacious materials and clearer definition on the surface of the sheet in contact with the mold.

Mechanical Forming

Pre-cutting negative forming (plug assisted) is combined to the base mold in order to obtain areas precise details on the surface of the sheet opposite that facing mold.

Pressure Diaphragm Forming

The sheet adheres to the mold through the use of an elastic diaphragm which is compressed by a high pressure fluid. Indispensable for extremely tenacious materials.

Twin Sheet Forming

Two sheets are formed simultaneously on two opposing half-molds, then welded together under high pressure. Cavities can also be created with materials of varying color and type.

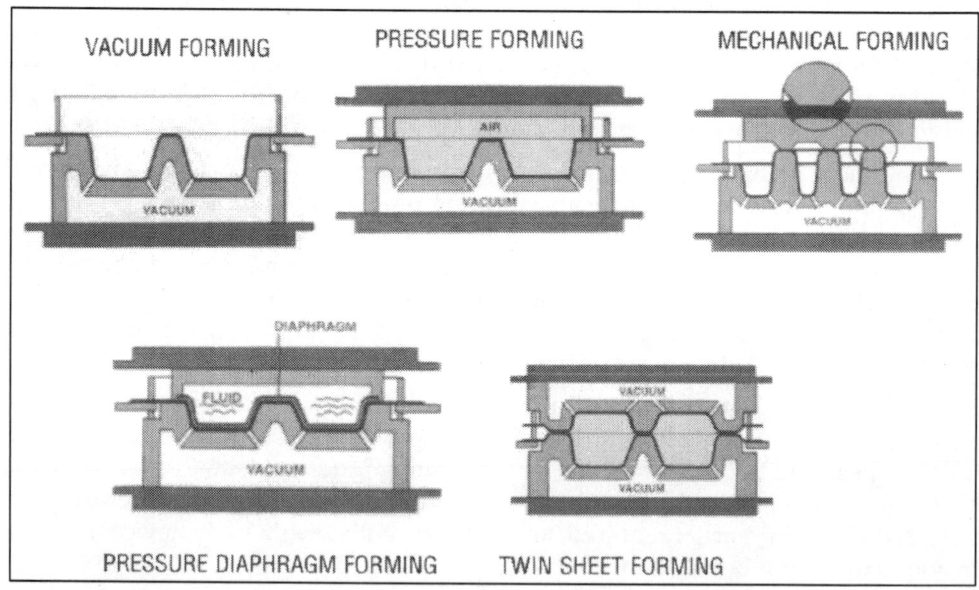

Benefits

Thermoforming technology provides far better advantages as compared to normal molding solutions. The reason behind this is that the thermoformed plastic products are manufactured at low pressure and hence possess superior physical properties. Using this technology it is possible to press multiple sheets at a time. Laminated as well as pressed sheets can be used in the process. The plastics can be given a desirable shape and details can be done on the molded side with much efficiency in a cost effective manner by using this technology. Larger plastic parts can be created with ease and efficiency by using the thermoforming technique. The detailing of the products is very easy and the end result after cooling showcases a high quality with smooth finish. Another major benefit of this technology is that it is much faster as compared to other molding processes and can produce high quality custom products at large scale in minimum time periods.

Thermoformed plastics have greatly influenced the products used in different industries and our daily lives. Most of the people prefer to purchase products that are manufactured by this process as they are highly durable.

Induction Sealing

Induction cap sealing occurs when a bottle, fitted with a plastic closure and aluminium foil liner, is placed underneath an induction sealing machine. The induction sealer transmits electromagnetic energy which is received by the aluminum foil liner causing it to heat up and weld itself across the container neck. The FDA recognizes induction sealing as an effective means of tamper evidence. This gives a sealed container, which

- Prevents product leakage
- Provides tamper evidence.
- Gives improved shelf life due to the excellent barrier properties of the foil and its airtight seal onto the container
- Improves pack presentation and customer acceptance.

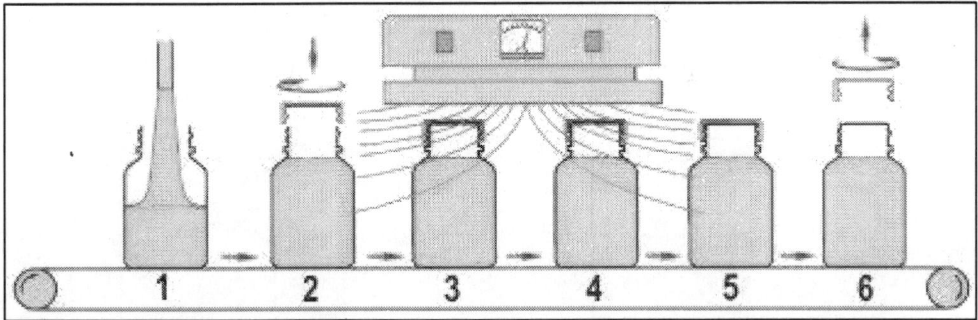

The bottle is filled at position 1 and the cap, complete with aluminum foil seal, is applied at position 2. The capped bottle is then passed under the induction sealing machine which causes the induction seal liner in the cap to heat up and bond to the container neck by position 5. When the bottle is later opened (6) the aluminum foil liner is released from the cap and left in place across the bottle neck.

Capless Induction Sealing

Can you Induction Seal without Using a Cap?

Yes, capless sealing is available. Induction sealing is often used to hermetically seal containers for freshness and to thwart tampering. But until recently, the process was limited to plastic containers with circular openings and threaded caps. Advanced sealing heads and more compact, higher-wattage power supplies, however, are making it possible to use induction for capless sealing (containers sealed prior to capping) and nonthreaded snap-caps, as well as metal containers and nonround container openings.

Capless induction sealing is fairly recent innovation. Normally, the sealing process requires heat and the pressure of the cap to weld the inner-seal to the container's lip. When there is no cap,

you need to apply pressure some other way. Some systems us an overhead belt to apply pressure. Other versions use pneumatics to apply pressure.

The Induction Sealing Machine

Induction sealers - sometimes known just as heat sealers or r.f. or high frequency generators - transmit an electro magnetic field which creates eddy currents in a metallic element, such as an aluminium foil induction innerseal, placed in the magnetic field.

This current heats the aluminium foil seal and its plastic coating bonds to the container neck. Induction sealing machines comprise a power source, a sealing head or coil and a cooling system. Manual hand held units are available for laboratory or low volume work but high power automatic cap sealers are usually mounted above conveyorised production lines.

Induction Sealing Machine Heads/Coils

The induction sealing machine's head, or coil, transmits the magnetic field. There are two main types, flat coils and tunnel coils.

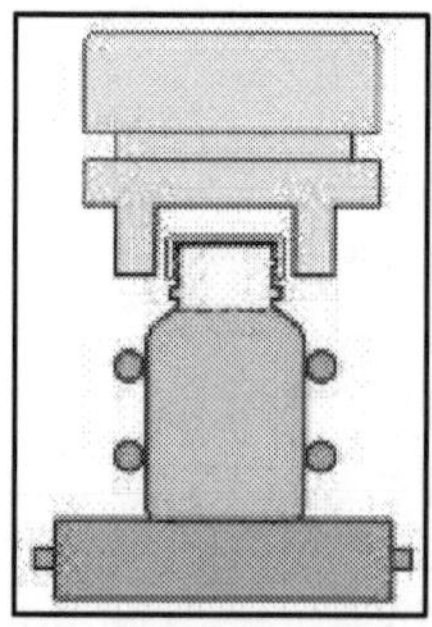

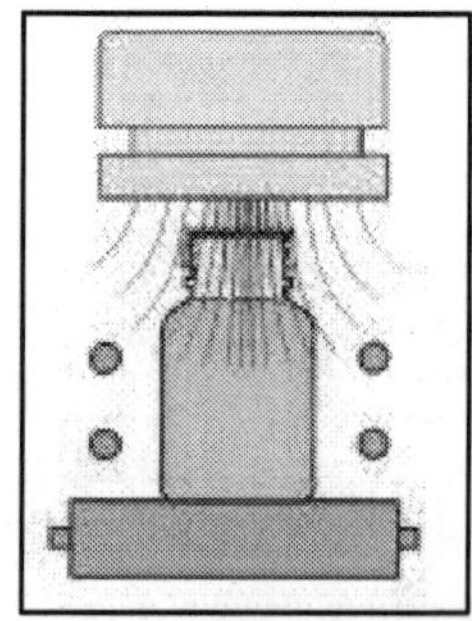

Tunnel Sealing Head/Coil **Flat Sealing Head/Coil**

The tunnel coil has the advantage that it locates the bottle centrally and often has a more uniform and deeper field because the coil windings can be incorporated in the sides around the cap. This makes it particularly suitable for closures where the induction seal liner is deep inside, *e.g.* sports caps and child resistant closures. Usually different sized tunnel coils are needed for different size caps. Flat coils are more universal and good for fillers using different cap sizes but special care must be taken to ensure that the cap is centrally positioned beneath the head. See requirements for a good seal below.

What Design Sealing Head is Best for a Particular Bottle Type?

The sealing head design is based on the point where the foil liner is placed in relation to the top of the closure. As mentioned previously, the magnetic field developed by the AC current applied to the sealing coil interacts with the foil disc. The distance which the magnetic field can propagate through air is very limited. Ideally, the foil disc should be approximately 1/8" to 1/4" away from the bottom of the sealing head. This area provides the maximum field strength. When the foil liner is further away, as in spout type closures or CRC containers, the field must be applied from the side of the closure. These types of sealing heads are referred to tunnel or channel designs. The locations of the components that develop the magnetic field are well below the top of the container, adjacent to the position of the foil liner in the closure.

Induction Seal Liners

Induction seal liners, or innerseals or wads, come in two main types, with and without a wax bonded backing liner these are often referred to as one piece and two piece:

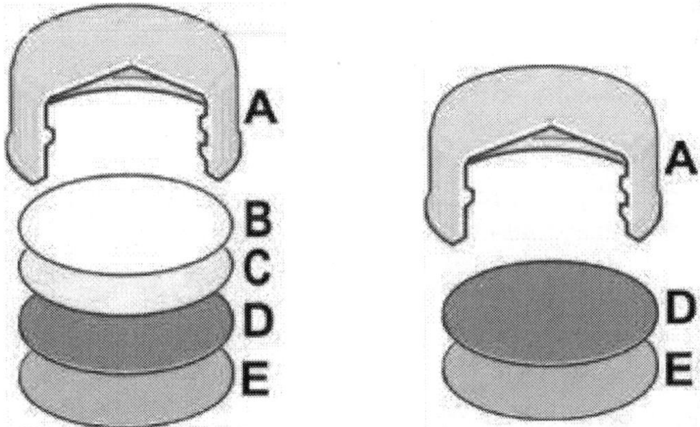

The facing of both types is similar, an aluminium film (D) laminated to a heat seal layer (E); the two piece type then has an additional backing liner or 'wad' (B) which is bonded to the facing layers D and E by a layer of 'wax' adhesive (C).

The one piece type of liner is removed from the cap entirely after the sealing process whereas the two piece leaves the backing liner (wad) in the cap for improved resealing after the initial opening.

The two piece liner has a 'wax' layer bonding the sealing layers to the backing board. The heat of the induction sealing process causes this wax layer to melt and it is absorbed into the backing board or other layer so that on cooling the two parts are separate.

The heat from the sealer causes the wax to melt progressively inwards from the edge. If correctly done the wax will completely free the liner facing, alternatively a small central area of unmelted wax may deliberately be left to retain the cap in place after sealing. This can avoid the need for subsequent cap re-tightening.

In all cases it is the heat seal layer which bonds to the container not the aluminium itself. The exact nature of this heat seal determines whether the induction liner will be peelable or full weld type and the types of container the foil will adhere to *e.g.* HDPE, PP, PET, glass etc.

Reasons that Induction Sealing may be Useful

Tamper Evidence

With the FDA regulations concerning tamper-resistant packaging, pharmaceutical packagers must find ways to comply as outlined in Sec. 450.500 Tamper-Resistant Packaging Requirements for Certain over-the-counter (OTC) Human Drug Products (CPG 7132a.17).

Induction sealing systems meet or exceed these government regulations. As stated in section 6 of Packaging Systems :

"...6. Container Mouth Inner Seals". Paper, thermal plastic, plastic film, foil, or a combination thereof, is sealed to the mouth of a container (e.g., bottle) under the cap. The seal must be torn or broken to open the container and remove the product. The seal cannot be removed and reapplied without leaving visible evidence of entry. Seals applied by heat induction to plastic containers appear to offer a higher degree of tamper-resistance than those that depend on an adhesive to create the bond..."

Leak Prevention/Protection

Some shipping companies require liquid chemical products to be sealed prior to shipping to prevent hazardous chemicals from spilling on other shipments.

Freshness

Induction sealing keeps unwanted pollutants from seeping into food products, and may assist in extending shelf life of certain products.

Pilferage Protection

Induction-sealed containers help prevent the product from being broken into by leaving a noticeable residue from the liner itself. Pharmaceutical companies purchase liners that will purposely leave liner film/foil residue on bottles. Food companies that use induction seals do not want the liner residue as it could potentially interfere with the product itself upon dispensing. They, in turn, put a notice on the product that it has been induction-sealed for their protection; letting the consumer know there was a liner on the plastic bottle prior to purchase.

What's New

Induction sealers generate heat, and manufacturers have traditionally cooled them with water. In the last 5 years, however, air cooling has gained favor thanks to improvements in power supply technology. Today, air-cooled power supply units can operate at speeds once only possible with water-cooled units. In addition, air-cooled units cost less that water-cooled ones and eliminate the hassle of maintaining a water recirculation system. While water-cooled systems have a small advantage in a few applications, air-cooled sealing systems will usually do the job.

Induction Seal Quality Tests

The simplest methods of testing bottle seal quality are the 'squeeze', 'stand on' or 'lay on side' leak tests. Full evaluation of cap seal quality is much more complex as other factors, such as foil degradation, long term product compatibility, wad stick etc. may be potential problems even though the seal passes the initial leak test.

Benefits of Induction V/s. Conduction Sealing

Conduction sealing requires a hard metal plate to make perfect contact with the container being sealed. Conduction sealing systems delay production time because of required system warm-up time. They also have complex temperature sensors and heaters. Unlike conduction sealing systems, induction sealing systems require very little power resources, delivers instant startup time, and its sealing head can conform to "out of specification" containers when sealing.

Applications

- Food and Dairy Products; Pharmaceuticals;
- Oil and Lubricants; Cosmetics;
- Agro chemicals;
- Paints, inks, dyes etc.;
- Laundry detergents/powders.

Symptoms of Under-heating	Lower Limit	Qualities of A good seal	Upper Limit	Symptoms of Over-heating
Leakage – no seal or incomplete seal		No leakage - complete circumferential seal		Possible leakage
Thin or incomplete 'witness mark' on foil		Good witness mark showing complete seal		Degradation of foil, from micro blisters to serious burning
Low peel strength		Correct Peel strength/weld strength		Difficult to Peel (peelable foils)
Little meltdown of neck		Some melt down of neck – depends on material		Excessive melt down of neck
Cap is difficult to remove (inadequate was melt on 2 piece liner)		Cap can be removed easily		Cap is loose on bottle due to excessive neck meltdown
		Backing wad (2 piece liner) in good condition		Scorching of Backing wad (2 piece liner). Burning smell
				Backing wad on 2 piece liner may bond to facing and/or come out of cap if hot melt adhesive is used
		Tab is free		Tab may stick to cap, bottle neck or itself
				Reduced shelf life and possible leakage or attack of alu. foil due to degradation

Conclusion

In order to achieve a proper seal,

- the foil liner must fully contact the lip of the container with sufficient torque,
- there must be material compatibility between the liner and the container,
- there must be sufficient exposure to the magnetic field to heat the foil to accomplish the seal,
- the power setting on the heat sealer,
- the distance between the cap and the heat sealer's coil, (This is critical as the strength of the field varies inversely with the distance from the source, doubling the distance reduces the field strength to one quarter)
- the conveyor speed which determines the dwell time of the foil seal under the sealing head.
- A failure at any point in the process can result in inconsistent or poor seals.

23

Packaging Line

Form Fill Seal Machines

Background

Proper packaging increases the quality of any product to a great extent. Every manufacturer knows this fact and takes great pains to pack their products in a proper way. Packaging prevents the product from getting damaged in transit too. This has made packaging industry a very prominent peripheral service in the market today. Along with this the demand for newer packaging material has also increased. For packaging these products one of the principal activities is filling. Filling is defined as transferring products to containers/packages for distribution. The demand for packaging machinery has also gone through a revolution. Packaging machines are of different kinds. The features of these machines differ along with the materials they use. Many types of packaging techniques are in vogue. Today, the various activities on the packaging line like making of the packs, filling the packs and sealing has been integrated to form the concept of FORM, FILL and SEAL machines.

Form Fill and Seal Concept

A machine that forms the pack, fills the product and seals the package is termed as Form Fill Seal (FFS) machines. These machineries have been broadly gained acceptance due to its applications in an extensive array of products. Form, fill and seal machines are packaging equipment that uses flexible, heat-sealable, plastic film to form packages that can be filled with a product and then sealed, and cut. Selecting form, fill and seal machines requires an analysis of specifications, features, and applications. Typically, fully automated machines list the rate, which is usually expressed in pieces per minute. Some automated machines include an integral feeder and a computer interface that links to a control network. Others are made from stainless steel for improved corrosion resistance. Aseptic or sanitary equipment is easy to clean and designed in a manner that inhibits the growth or presence of pathogenic microrganisms. Portable machinery is lightweight or includes wheels or casters for ease of movement. In terms of applications, form, fill and seal machines are used widely in the automotive, chemical, medical, and pharmaceutical industries. They are also used to process food, beverages, cosmetics, electronics, semiconductors, stationary, tobacco, and military products.

The main types are **vertical form fill seal** (**VFFS**) and **horizontal form fill seal** (**HFFS**) machines often applied in areas of horizontal versions of flow-wrappers, sachet machines, blister pack machines, four side seal machines and thermoform fill and seal machines.; in both cases packaging material is fed off a roll, shaped, and sealed. The bags / packs are then filled, sealed and separated.

FFS machines can be embedded with highly sophisticated features like detectors, computer interfaces, and control networks. Greater speed and versatility are the major benefits of FFS systems for user companies. The capability of creating virtually any size or shape has enhanced its relevance in various sectors like FMCG or otherwise. The wide diversity and the multiplicity of FFS machines employ a wide range of material types and are used across numerous markets including food, drinks, cosmetics, electronics, stationary, tobacco, chemical, medical, and pharmaceuticals. The different packaging system applicable are : lower reel flow-wrapper, Bags and Pillow packs, Bottles / vials, Cartons, Pots trays and blister, Sachets and Envelopes and Sacks and tubular sack Bags.

Vertical Form Fill Seal (VFFS)

In this case, the flexible packaging material forms a tube, contains the product and is sealed in a sequence of operations, whilst the film is transported vertically downwards.

Vertical Form Fill and Seal Machine for Cartonboard

A vertically operating form fill and seal machine which uses a cartonboard laminate, which is formed, filled with product and sealed to produce a pack resembling a carton.

Tetra Pak carton for milks and juices is a variant of a carton VFFS pack.

Mandrel Flexible Package Form Fill and Seal Machine

A packaging machine which forms packs from a reel of flexible material, on one or a number of mandrels, before filling the packs with product and sealing their tops within the machine. Packs are formed using flexible packaging material on one or more mandrel to form the pack which is then filled and sealed within the machine.

Uses: Traditionally used for flour and sugar.

Tubular Form Fill Seal (TFFS)

Tubular Form Fill Seal covers two main types that cover larger bags and sacks:

Tubular Bag Form Fill and Seal Machine

Here a reel of lay flat tubular flexible packaging film forms a bag in which the product is filled and sealed within the machine.

Tubular Sack Form Fill and Seal Machine

Here a reel of lay flat tubular flexible packaging film forms a sack. The sack is then filled with product and sealed either within the machine or by separate machines.

Sachet Form Fill Seal

Here the operations remaining the same, three sided or four sided pouches can be made.

Three main types are:

Edge Sealing Machine

Horizontally operated form fill seal machine in which the product is placed on a horizontal web of film before being sealed on 3 or 4 sides to an upper web of film. Machines may have one or two reels of material and can produce one or more lanes of packs.

Horizontal Sachet FFS

Packs are formed and sealed on 2 or 3 sides and filled vertically with the product before the remaining side is sealed whilst the film web is moving horizontally but the pack remains vertical.

Uses: liquids, free flowing powders and wipes such as hand cleaners, cosmetics etc.

Vertical Sachet FFS

Uses one or more webs of film to form a pack which is filled vertically and sealed

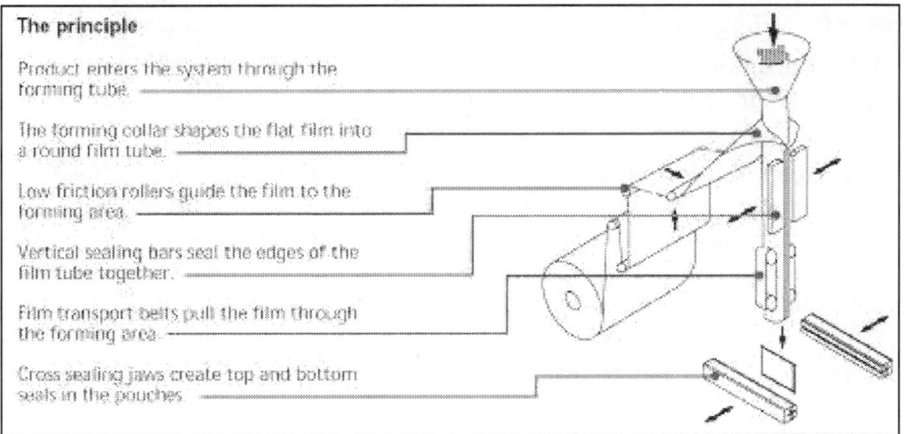

to produce a 3 or 4 sided sealed sachet. Can have one or two webs of film and several lanes of packs.

Uses: tea bags and pack liquids and free flowing powders and solids.

Thermoform Fill Seal

A form, fill and seal machine in which a web of thermoformable material is heated and formed with pressure and / or vacuum, before being filled vertically with product, sealed with a top film or magazine fed lid and finally cut to produce individual packs. Machines can produce one or more lanes of packs and may incorporate equipment to evacuate packages before they are sealed.

Blister Thermoform Fill Seal

The most common form of thermoform fill seal used almost exclusively pharma pills and capsules.

Deep Draw Form Fill and Seal Machine

Packaging machines which use deformable or thermoformable packaging material, to form a package which is then filled and sealed in a sequence of operations to form a deep drawn pack.

Cold Form Fill and Seal Machine

A form, fill and seal machine in which a web of deformable material is formed under pressure in a die press, before being filled vertically with product sealed, with a top film or magazine fed lid, and finally cut to produce individual packs. Machines can produce one or more lanes of packs. Uses: single serve jam and preserves.

Filling and Dosing Machines

Filling and dosing machines are packaging machines which measure out a product from a bulk supply by some predetermined value, *e.g.* volume, level in a container, mass or count.

The **filling** method used is influenced largely by the nature of the product *e.g.* liquid, gas, piece goods, powder, free flowing solids or sticky paste, but also by the measure for selling the product *e.g.* by weight, by volume or by count.

Filling machines may comprise of one or a number of dosing devices that may be arranged with or without a mechanism to control containers or packages as they are filled.

The two most common arrangements of fillers with multiple filling heads are 'in line' where the filling heads are fixed and containers are moved under the heads and then removed when filled in an intermittent process, and 'Rotary' where several filling heads are mounted on a rotating carousel and containers are fed onto the carousel, filled and then discharged continuously.

Volumetric Fillers

Volumetric Piston Filler

Filling mechanism which measures out a product, usually liquid, gas or paste, using a reciprocating piston of predetermined volume. Fillers of this type are suitable for accurate filling of products at both low and medium speeds.

Displacement Filler

Filling mechanism which measures out a product by displacing a predetermined volume of product with a loose fitting piston. Displacement fillers are slightly less accurate than piston fillers but are used in situations where it is important not to damage the solids in the product.

Aerosol Gassing Machine

A volumetric piston filling machine designed specifically to fill propellant gas into an aerosol can or canister. Fillers of this type are suitable for accurate filling at both low and medium speeds.

Rotating Chamber Filling Machine

Filling mechanism which measures a product, usually liquid gas or paste, using a metering pump that operates for a predetermined number of cycles.

Flow Meter Filler

Filling mechanism which measures out a product, usually liquid, using a flow meter. If based on an inductive flow-meter these machines are restricted to use with water based liquids but have the benefit that they are easy to clean and are tolerant to included solids.

Auger Filler

Filling mechanism which measures out a product, usually powder or free flowing solids, using an auger which is rotated for a predetermined number of revolutions in a conical hopper to discharge the required volume of product. The main benefit of these machines is their ability to control dust during the filling operation and are therefore used extensively for powders and dusty free flowing solids. To compensate for changes in the bulk density of the product, auger filler are frequently used in conjunction with a weighing instrument like a checkweigher. Fillers of this type are suitable for filling products at both low and medium speeds.

Volumetric Cup Filling Machine

A filling which measures out a product, usually free-flowing solids or powder, in a cup of predetermined volume. These machines are typically used for products like rice or frozen peas which do not generate dust. Fillers of this type are suitable for accurate filling of products at both low and medium speeds.

Timed Flow Filler

Filling mechanism that measures a product, usually liquid or powder, by controlling the product flow duration to a predetermined value. This indirect method of achieving a volumetric fill has the benefit of being relatively low cost and easy to clean. Fillers of this type are used in situations where a single machine is required to fill a wide range of product volumes. Timed flow fillers are generally used for low speeds.

Gravimetric Filling Machines

Net Weighing Machine

Filling mechanism which measures out a predetermined mass of product, usually free flowing solids, before dispensing it as a fill. These machines can weigh products as diverse as sand, vegetables and breakfast cereals. Fillers of this type are suitable for weighing products at both low and medium speeds.

Selective Combination Weighing Machine

Also known as a Multi-Head Weighing machine, this is a nett weighing machine with multiple weighing units that computes an appropriate combination of loads to achieve a predetermined mass, before discharging them together as a fill. (See also Form Fill Seal and Inspection and Weighing Pages). Weighers of this type are suitable for accurate filling of products like potato crisps or salads at both medium and high speeds.

Gross Weighing Machine

A filling sometimes known as a weigh filler which measures out a predetermined mass of product, which may be liquid, powder, gas or solids, directly into the package which is resting on a weighing instrument which controls the filling operation.

This method of weighing is used for products as diverse as jars of instant coffee, litre bottles of oil and 200 litre drums of chemicals. Gross weighing machine are suitable for accurate filling of products at both low and medium speeds.

Count Filling Machine

Packaging machine used to measure solids according to a predetermined count. A machine used commonly to pack pills into small rigid containers, or in hardware packing for small DIY items such as screws and nuts.

The most common forms of inspection machines are check weighing machines and metal detecting machines but the range of inspection equipment available has grown hugely in recent years driven by the need to automate production and to remove the people from the line who as well as carrying out manual tasks also inspected products and packs visually. In the pharmaceutical industry a range of inspection systems have also been developed to check that batch codes are present and legible and to check that every packaging component is correct for the product being packed. Inspection machines are used in all end user sectors, including food, beverage, pharmaceutical, toiletries and cosmetics, and household chemicals. Some kinds of inspection are more common in some sectors than others or even unique to a particular industry (for example missing or broken pill detectors in the pharmaceutical sector).

Conclusion

The concept of FFS encompasses the pouches, tetra packs, thermoformed trays packs like jams, buuter and in a way the lined cartons. The main advantage is high speed and high quality production. With revolution of flexible and semi rigid packaging materials, the future of FFS systems will be on the increase. The future developments would be to focus for higher productivity like high speed, multi lane machines. It is imperative that as the FFS technology takes a leap the packaging material to be used on these machines will also have to match up with the precision and quality. In summary there would an up gradation of the packaging material quality vis a vis the packed products.

24

Packaging for Export

Packaging Graphic Design for Exports

The "Added - Value" Concept

Most developing countries have literally centuries – long experience in exporting their basic Commodities, such as tea, coffee,spices,cocoa,rice,sugar, cotton or bananas. Most of these products are traditionally exported in bulk, to be repacked in consumer units by the importers / buyers in industrialized target markets.

The supplying countries, however, will not long be content to export their products in bulk as raw materials, or even in a semi – processed form, but will, in the future, strive towards an increased local processing of their indigenous raw materials. There is, today, much discussion about the necessity to upgrade export earnings so that the added value, resulting from processing, may be kept in the producing country. This "added – value" concept will proceed slowly but steadily, and will involve a lot of difficulties in technology, trade policies and marketing methods. There is, however, no doubt that this process has already started and will be accelerated during the next few years.

New, Non-Traditional Export Products

In addition to the increased interest towards consumer – packing of traditional commodities, much effort is spent on developing and introducing new products through new exporters on new markets. The trend towards both the local processing of traditional commodities for export, and the introduction of new, non – traditional, consumer – packed products on the world market, will mean that the exporters in developing countries will have to pay close and increased attention to customer requirements in sophisticated industrialized consumers in the target markets, using own brand names, introducing new types of products, and maintaining competitive packaging designs and high quality levels. This will also bring about new and, often, difficult problems for who are often unfamiliar with, for instance, the requirements of the supermarket distribution systems, or who have difficulties in understanding the basic concepts of package design on their target markets.

Domestic versus Export Packaging Design

Highly promotional packaging design, an active ingredient in competitive marketing, may not always be so important for the distribution of goods in the domestic markets of developing countries where the producers are locally well known and are operating in a fairly easy monopoly or oligopoly business situation. However, when export of the same products is considered, these products enter into the world markets facing severe competition from a FOR large number of experienced rival producers. Now promotional packaging becomes a vital element in the marketing strategy. Unfortunately, customer requirements as regards export packaging are often forgotten or neglected by the exporters in developing countries. People involved in the development of packaging seem to have great difficulty in assessing the difference between the design of domestic packaging and the design of export packages. This can usually be attributed to a lack of knowledge of the needs and requirements of their customers abroad.

Developing and Developed Target Markets

A very important and recent concept in the promotion of exports from developing countries is trade between developing countries themselves. As an example, in Asia, the export of consumer – packed goods to the oil – rich Arab countries can be mentioned. Other examples would be import – oriented countries like Singapore and Hong Kong. A widely spread misconception is that these markets do not require high quality, sophisticated packaging. However, one should not forget that rival producers / competitors in industrialized countries are equally interested in these markets and, consequently, the products will find themselves beside the world's most sophisticated, high quality packages / designs on the shelves of the supermarkets.

At this stage it is, perhaps, necessary to point out that this presentation does not deal with export packages / labels for consumption by the often large groups pf migrant populations in industrialized target markets. Due to the limited quantities of products available for export, many producers in developing countries are basing their marketing on the purchasing potential of their countrymen abroad. In this case, the distribution is handled through specialized channels and the same, domestic packages, modified only to meet the legal requirements of the importing country, can be used.

Another short cut in dealing with export packaging problems is to make an agreement to pack under an importer's / distributor's brand. In some instances the customer will provide his own packaging materials and designs. However, packing under contract should only be seen as a temporary measure since this method usually results in the lowest possible price for the producer. There are definite advantages also – larger quantities, only one buyer, less marketing efforts and costs, etc. The implications of this paper, which concentrates on the development of consumer packs for export under own brand names, and a direct contact with the distribution systems, primarily in industrialized target markets.

The Current Position

Based upon ITC's experience and observations during the past nine years, the current position of the graphic/promotional designs used for consumer packed products, exported from developing countries in all parts of the world, leaves much to be desired and opens up an enormous field of work for improvements in the future. In summary, the most important problems seem to be:

- Incorrect positioning of the designs on foreign markets;
- The lack of a systematic family appeal between the different products of the same producer/exporter;
- The quality of the execution of designs (printing) on export packages/labels is too low;
- Old – fashioned and inconsistent use of typography;
- Frequent violations of the packaging – related regulations in the foreign markets;
- Unnecessarily high packaging costs, mainly as a result of the exporter's lack of technical/commercial know – how;
- Difficulties in introducing joint export marketing strategies between producers/packers of related products.

These seven problem areas will be further elaborated on below:

1. Positioning

The most important background concept to all development of promotional package/label designs is how the product should be positioned on the market. This decision has to form an integral part of the overall marketing strategy, and must be taken by top management already at the initial planning stage.

The first question to ask is general in nature: should the layout of the package/label design be traditional – only slightly different from competitors' designs, or should it be radically different, seeking high attention/a contrasting position on the supermarket shelf? The main question, however, is more important and, at the same time, more complicated. There are three three major concepts which can be expressed in the overall layout/positioning of the package/label design:

a) The brand : Presenting the corporate brand mark, usually expressed as a logotype, or the product brand name (*e.g.* "Tropicana" or " Ocean Queen", etc.);

b) The product: Presented in words or illustrations and, whenever possible, emphasizing a USP (Unique Selling Proposition), *i.e.* a specific advantage, or strong sales argument in favour of the product;

c) The target : Presenting the target consumer, *e.g.* men / women / children, the whole family as a group, ethnic groups of people, etc., or the consuming situation, *e.g.* breakfast meal, picnic, Kitchen cleaning, etc.

The totality of these three concepts can be expressed in one sentence:

WHO (brand) SELLS WHAT (product) TO WHOM (target)?

The important policy decision, in this context, is to decide how the package / label design should be positioned on the market – should the layout concentrate on only one of these concepts or should it consist of a combination of two or all three of them (and in what proportions)?

Current package / label designs in developing countries concentrate heavily on the brand concept. This strategy will be effective only if the package design can be backed up by extensive advertising or other forms of supporting promotion / publicity. This, however, is very seldom possible for exporters in developing countries due to the lack of financial resources and, therefore, it would seem to be more effective to concentrate on the other two concepts – product and / or target – in the positioning of the product on foreign markets.

The reasons for the current concentration on brands in package / label designs by producers in developing countries are quite natural. Often in a monopoly or oligopoly situation, the brand on the local market is important. It is, therefore, very difficult for the producer to realize that his brand, his factory or his family traditions do not mean a thing to the consumer in a country far away – the housewife in Sweden or France is not the least interested in the brand, she is only interested in the product and its advantages to her and her family. Since she will not learn about the product through advertisements or other publicity, she will make the purchase based upon an impulse, standing in front of the supermarket shelf. The package / label must, therefore, be able to:

- Visually attract the attention of the consumer and arouse her interest. Since there are so many competing products on the supermarket shelf, this must be done in a fraction of a second.
- Create consumer confidence in the product – the package has now been picked up for closer examination. Dented lack of product information are example of details which will kill the sale at this stage.
- Be clearly identifiable among other products on the shelf so that the consumer can easily find the package again when she comes back for a repurchase.

It goes without saying that, first of all, the quality of the product has to be acceptable to the consumer – packaging design cannot sell an inferior product.

2. Family Appeal

A very common mistake in package / label designs in developing countries is the lack of co-ordination among the designs for the different products manufactured by the same producer. Sometimes one can find as many as 5 -10 different versions of the same brand mark (often in different colours) and the general layout often lack conformity from one product to another. This usually happens when new items are added to a product line – a new sales manager also likes to add a new look to the existing designs and, after a few years, the packages/labels lose all their "family look". This considerably reduces the total impact of the product line on the supermarket shelves in the target markets.

Even if the emphasis in positioning is not put on the brand, it is very important that all the packages are clearly seen as having the same identity. This is particularly true when there is no advertising or other promotion for the products. If the producer likes to differentiate between his various background for fruit preserves, green for juices, yellow for vegetables, etc). This is common practice *e.g.* for medicine and pharmaceuticals. The important thing is that the general layout does not change, and that the brand name is always presented in the same way, scale, position and colour, on the packages/labels in order to reinforce the impact of the product family and the corporate identity.

3. Execution of Designs

As noted earlier, only a good quality execution of the package/label design will create the essential confidence towards the product in the mind of the consumer. In many cases this is related to the technical aspects of the package – it should look neat and carefully put together. Rusty or dented cans, scuffed or torn labels, defective or badly executed heat seals in plastic packages, wrinkled wrappers, roughly or improperly die – cut folding cartons made out of low quality paperboard, the poor materials used and the lack of the most common technical deficiencies found in consumer packaging from developing countries. Our main concern, in this context, however, is the graphic design of the printed package/label, and how this is executed in the final perfectly able to produce adequate quality printed packages and labels. Then why are so many export packages from developing countries badly printed? Here are some of the reasons:

- Management does not appreciate the need for quality in print execution and places the orders with small, cheap but incompetent, printers;
- The availability of adequate quality paper and carton is restricted, sometimes through government – imposed import policies. This is also very often true for essential, auxiliary materials, such as photographic film, printing plates and inks, varnishes, etc.
- The quality of the original art work material, particularly the photographs (diapositives), to be used for the illustrations is of low quality or not suitable for reproduction in print. Good illustrations cannot be printed from snapshots taken by amateurs – the taking of *e.g.* product pictures should be left to professional photographers. Naturalistic drawings of *e.g.* food products are very difficult to make and seldom come out well in the final print. If photographic separation of colours cannot be used, it is better to base the illustrations on line drawings or other types of design techniques which are easier to reproduce. Large areas of solid colours should also be

avoided since they will only create drying/blocking problems for the printer, and unnecessary difficulties in maintaining a consistent colour coverage/ shade.

4. Typography

A small, but important detail to consider in a well – balanced graphic designs for packages/labels is the choice of typography (style of letters). Two problems can be observed in many of the graphic designs currently used in developing countries:

- The typography is old – fashioned. To a certain extent, packaging typography is a question of fashion in the same way as colours, shapes and product designs. There is really no excuse for using out – of – date typography, particularly since the type – setting can easily be done with self – adhesive or transfer letters (such as Letraset), which are both inexpensive and effective;
- The typography is inconsistent. This means that several different types of letters have been used in various parts of the graphic layout. The overall impression usually be comes unbalanced and confusing. The use of different types of lettering in the same layout requires long experience and a very good conception of typography.

5. Regulations

One of the most important problems in packaging for international trade is the necessity to observe all of the many, and often, complex laws and regulations are sometimes seen as consciously created "invisible trade barriers", particularly aimed against low – price imports from developing countries. This is, generally speaking, not true. The regulations reflect genuine interest on the part of consumer organizations and governments to protect with consumer packaging. Unfortunately, however, many such regulations are rather badly conceived and have often been introduced without proper consultation with all parties involved, *e.g.* the packaging industry. Consequently, the laws and regulations related to packaging are currently seen as one of the major problems for both the packaging and the packaging industries, difficult to comply with even for producers in industrialized countries.

Regulations related to packaging are basically of three types:

- Those related to the protection of the consumer's health. They might refer to the migration of harmful substances from the package itself into the packed foodstuff, or they might refer to child – proof closures, technical specifications, and mandatory warning texts to be printed on packages containing hazardous products, etc;
- Those related to fair trade practices.They might state what product designations may or may not be used; what kind mandatory information is to be given by the producer as to the content of the pack, the ingredients used in the product; and they might also ban the use of deceptive product illustrations, unwarranted claims and unnecessarily large package sizes, etc;

- The third category, which does not take the form of legally enforced regulations, consists of the voluntary standards and practices used by the trade in the target markets. The UPC/EAN bar coding symbols and standardized package sizes, based upon the module of 400 x 600 mm, are examples in this context. However, the implications of such voluntary regulations for the exporter are as important as any mandatory, legal regulation – non –compliance, in practice, will simply mean no sale.

The most important laws and regulations related to retail packaging of consumer goods are the Council Directives, adapted by the European Economic Community (EEC), and the Fair Packaging and labeling Act in the United States.

As to the Question of whether current packages / labels from developing countries comply with the existing laws and regulations in the industrialized target markets, the answer, unfortunately, has to be negative. The fact that many of these packages are still able to enter the target markets is merely a matter of insufficient import controls, and lack of legal enforcement of the regulations. It is expected that such controls will be made more efficient in the future. Therefore, it is expected that such controls will be made more efficient in the future. Therefore, it is of utmost importance for exporters in developing countries to keep themselves informed about packaging – related laws and regulations in the prospective target markets, and to adapt their package and graphic designs to these requirements. The International Trade Centre will compile and public guidelines on this subject during the first half of 1982.

6. Export Packaging Costs

In many cases, the costs of export packaging in developing countries can be up to 40 – 50 per cent of the net FOB export price, even for standard low – price commodities. Every saving that can be made, without jeopardizing the technical or commercial performance of the package, will improve the exporter's competitive position on the world market. In order to be able to critically analyse current packaging practices and to find more economical alternatives, the exporter / producer himself must have a good knowledge about packaging technology and between the user and the supplier of packages and packaging materials. The present situation in most developing countries leaves much to be desired in this respect. This is an area where a country's packaging institution can make a very important contribution to the national economy by initiating and implementing systematic, long – range training programmes in the field of packaging technology and promotion.

The International Trade Centre has made a list of important points to consider in improving packaging economy and avoiding unnecessary waste. This list is available, on request, from the ITC secretariat in Geneva. One point, selected from this list and relevant to the subject of today's discussion, deals with the number of printing colours.

7. Exporters' Joint Actions

It is amazing to observe how difficult it is for exporters in developing countries to collaborate, and pool their scarce resources in order to achieve a more substantial marketing impact and penetration into foreign target markets. In the field of export packaging, there are many ways of establishing a better packaging economy, and a more consolidated presentation of the products on foreign markets, through collaboration between exporters. The apparent economy of using standardized sizes, is one example in this context (e.g. 5 1bs. frozen shrimp cartons), or the joint procurement of larger quantities of packaging materials. Contract packing – a concept which seems to be impossible to introduce in developing countries – would provide particularly small exporters with good possibilities to upgrade their packaging technology to a level that would meet the requirements on the target markets. Joint promotional efforts, such as, for instance, using national standard printing on boxes for fresh fruit and vegetables, cut flowers, handicrafts, seafood,etc. would be much more effective and more widely observed in the market places of foreign countries. National slogans, printed on the packages, could be, for instance: "Beautiful fresh flowers directly by air from the tropical gardens of Sri Lanka ", " Delicious Malaysian shrimps" or " Quality handicrafts from the exotic Philippine Island", etc.

Three common misconceptions can be recognized, in this respect, as typical for most developing countries:

- Producers / exporters consider their next door neighbours to be their worst competitors when, infact, the most severe competition comes from producers in other countries;
- There is a constant fear of disclosing exported quantities, new product designs, names of customers, etc., when in fact, these data are already know or can easily be acquired, if needed. Joint purchasing schemes for packaging materials and contrast packing stations are also possible to set up on a confidential basis, if necessary;
- Each exporter / producer, however small, likes to see his own name in large print on the packages, whereas the consumer is totally uninterested in knowing from which garden, fishery, farm, or village the product comes from. A consolidated country brand, or a quality certificate from a joint export organization, would be much more important to him.

8. Checklist for Consumer Packages for Export

In order to assist exporters in developing countries to analyse their current export packages / labels for consumer – packed products, and to plan for new, improved and more effective designs, a checklist has been worked out and is attached to this paper. Whereas a checklist, as such, does not solve any problems, it helps in identifying the types of problems faced, and also serves as a list of all the details which should be observed in package development in order to prevent some, maybe essential, point from being overlooked.

25

Cost Reduction

Cost Reduction in Packaging

STUDY 1 - GLASS TO PET BOTTLE

Warner-Lambert Company

182 Tabor Road

Morris Plains, NJ 07950

Summary

Switching from a glass bottle with corrugated and paper overwrap to a plastic bottle without additional packaging, Warner-Lambert eliminates 20 million pounds per year of packaging for the Listerine mouthwash product, a 52 per cent weight reduction.

Action

Warner Lambert manufactures health care and consumer products. Listerine, which has been produced since 1906, is regulated by the federal Food and Drug Administration because of its claim to kill bacteria leading to gingivitis and plaque.

Consumer complaints to WL about unnecessary packaging of Listerine led the company to examine the feasibility of making a change. The process required a sizeable capital investment, complex coordination among departments, long lead time, and patience. All the affected parties were involved: manufacturing, marketing, legal, purchasing, engineering, environmental, packaging technology.

Among various factors considered were the following:

1. Redesign of the bottle in such a way as to deter competitors from copying the distinctive look of Listerine.
2. Development of a child-resistant cap closure that's easy for others to remove.
3. Shelf-life testing.
4. Drop-test proof.
5. Consumer acceptance—no drop in marketshare.
6. Safety. The old container presented the messy risk of a glass bottle filled with an alcohol containing liquid. What were the risks associated with a new container?
7. New machinery. A single plant produces Listerine for the entire market.
8. Distribution, transportation—savings expected to accrue from reduced product weight.
9. Overall cost saving / payback.

The new Listerine bottle is manufactured of polyethylene terephthalate (PET) and has been well accepted by consumers, the company reports. Interestingly, WL tried to change ots packaging this way in the early 1970s but was rebuffed. Focus group research at the time found that consumers felt such a packaging change signaled a cheapening of the product. Today, such waste prevention initiatives are applauded as evidence of a company's environmental sensitivity.Payback

WL will not reveal the overall cost of this packaging change. But the company is pleased with the change and expects to recover its investment in full. As one company official observes, a 52 per cent reduction in packaging weight has to be significant.

WL has considered packaging changes for other products, with mixed results. For example, it was found that Caladryl, an anti-itch lotion for poison ivy, could not be packaged in PET (the most recycled plastic) because the lotion interacted with the plastic. Hence the bottle remained in a polypropylene container.

CASE STUDY 2 - USE OF INCOMING MATERIALS FOR OUT GOING PACKAGES Schumacher Electric Corp.

513 N. Melville St., P.O. Box 39

Rensselaer, IN 47978

Summary

By using incoming cartons as feedstock to manufacture its own corrugated packaging for outbound freight, the company saves at least $60,000 a year in outside corrugated purchases and reduces production of baled corrugated for recycling by 16 tons a year, or about 70 per cent.

Action

Schumacher designs and manufactures electrical transformers for power, audio, and other applications. The products are small but heavy and require very secure packaging in corrugated cartons. Smaller units are packed in bulk quantities. To prevent damage during shipment to customers, the transformers are cushioned top, bottom, and sides with strips and rectangles of corrugated cut to special dimensions.

Prior to 1994, the company purchased these corrugated packaging materials from outside suppliers. Also, like many manufacturers, the company reduced its waste hauling expense by baling and recycling the corrugated cartons that arrived daily with incoming freight. One day, management observed the connection between these two activities; or, as manager Loren Snow summarized it, "Never buy something you're throwing away."

Schumacher began to manufacture its own corrugated packaging. Space was available near the baler, where incoming (used) cartons were brought to be baled from throughout the 100,000-sq.- ft. plant. Under the new procedure, only scraps and torn or dirty cartons go in the baler. Reusable cartons (estimated at 90 per cent of the total received with incoming freight) are flattened and stacked in neat piles. As needed, a small stack of flattened cartons (7 to 10) are lifted onto the bandsaw table and, guided by jigs, cut by the baler operator to dimension size for various use as cushions, shelves, and partitions inside cartons containing outgoing freight.

Payback

During the last full year that Schumacher purchased corrugated packing parts from outside suppliers, actual purchases exceeded $60,000. Since the conversion to inside manufacture of packaging parts required no capital outlay (bandsaws were available free and the feedstock remains free), there was no payback period—the reduction in outside cost was fully available at once.

Additional Waste Prevented

1. Storage space approximating 2,000 sq. ft. that had been required for inventory of custom cardboard pieces from outside suppliers.
2. Management believes that significant savings accrue whenever products purchased outside are manufactured in-house, in addition to the obvious reduction in vendor costs. "You avoid all kinds of paperwork," Mr. Snow observes. There is no need to maintain parts numbers, for example; no need to generate purchase orders, maintain inventory records, and handle all the paperwork associated with paying vendors. Although the company has not calculated these savings in the present case, Mr. Snow believes the reduced cost of clerical and supervisory time easily offsets the cost of the labor to produce a usable product from "scrap" corrugated.

CASE STUDY 3 - CONVERT FIVE -PIECE PARTITION BY ONE PIECE UNIT Jefferson Smurfit Corporation

1401 W. Ellerman Litchfield, IL 62056

Summary

By replacing a five-piece case partition with a lighter weight, one-piece unit, the company has reduced material usage about 20 per cent while increasing stacking strength.

Action

Jefferson Smurfit Corporation (JSC) is the third largest U.S. manufacturer of chipboard partitions, used to compartmentalize cardboard cartons of bottled goods. JSC also makes the largest multicell partition—one with 36 or more cells, mostly used for pharmaceuticals, cosmetics, and other small items. However, JSC was the smallest producer of the commodity partition—the 6-, 12- and 24-cell partitions used extensively in the food and beverage industry.To become more competitive in the commodity partition market, JSC decided to produce a new 12-cell partition. After 5 years of development, the company introduced its new "Q Stack" partition. The glued, one-piece unit is manufactured of.037 chipboard, made of 100 per cent recycled content. The new 12-cell partition replaces a slotted five-piece partition manufactured of.047 chipboard rollstock that was 20 per cent heavier.

Although the new, one-piece unit is lighter in weight, it is stronger than the old partition. This is due primarily to the fact that it is a solid piece. Tests show that the.037 Q Stack partition has greater stacking strength than 175# test, B-flute corrugated cardboard commonly used to manufacture slotted partitions, and is comparable in strength to 150#, C-flute corrugated. Users also note that the old, slotted partition could shift and fall out of the shipping case, interrupting operations.

Q Stack is produced in six calipers from.028 to.040, in cell sizes from 1-3/4 to 4 inches, and heights from 5 to 15 inches. JSC lists the user benefits as follows:

1. Increased stacking strength
2. Improved load-bearing strength
3. Better sturdiness
4. Easier to insert into the corrugated shipping case
5. Reduced case, partition, and product damage
6. Reduced package weight

JSC says it has benefited from the new partition by:

1. Reducing material requirements by 20 per cent
2. Reducing storage requirements
3. Reducing production costs by 10 per cent
4. Reducing production scrap rate Payback

JSC made a significant investment in R and D to perfect the Q Stack and bring it to market. The company expects to recover its development costs over time with increased sales of the innovative new partition. "There has never been anything like this available before," a JSC representative said. "This allows cost savings for our customers." Thus far the largest users have been glass bottle manufacturers.

CASE STUDY 4 - VARIETY REDUCTION IN SHIPPER SIZES

Xerox Corporation, 800 Phillips Road, Bldg. 205-99P, Webster, NY 14580

Summary

By converting from a system that used many different size one-way shipping containers to one that relies on nine standardized reusable corrugated package sizes, Xerox factories and suppliers worldwide can reuse the same boxes, thus diverting a large volume of corrugated packaging and wooden pallets from disposal and cutting storage and shipping costs.

Action

Xerox Corporation manufactures and sells copiers and other office products worldwide. In recent years, competition has led Xerox to reevaluate operations and costs, including its supplier packaging program. Work began at the Webster, New York, plant. In the past, this facility received component parts from more than 400 suppliers, each part packaged in its own, unique box. This resulted in the use of thousands of different types of boxes and as many as 24 different pallet sizes. To gain better control, Xerox developed a box reuse program (called the 88p311 Supplier Packaging Program). The company brought together packaging engineers from Xerox and its international suppliers to achieve a consensus on box style. Other participants in the planning process were quality control engineers, buyers, line engineers, assembly line workers, suppliers, and box makers. The program received final approval from senior manufacturing management.

The key features of the system: (1) Nine standard corrugated cardboard box sizes and two standard wooden pallet sizes have been adopted; suppliers are required to use them. A "Supplier Packaging Agreement Form" specifies how parts are to be delivered to Xerox facilities and describes under what circumstances exemptions are allowed. (2) A third-party handler manages the collection, sorting, and reselling of empty containers, eliminating the need to return containers to their point of origin. This is an "open loop" system.

Boxes and pallets can be used at any Xerox facility. They are designed to fit directly into designated positions on the assembly line, compatible with just-in-time delivery. As incoming shipments of parts are received and used, boxes are collapsed and stacked. The company then either reuses them itself to ship parts to other Xerox facilities or repair centers, or it sends boxes to the third-party handlers (one on each coast and one in the Midwest) who sort and resell the boxes to Xerox suppliers.

Payback

Before adopting the new system, Xerox spent over $500,000 a year at Webster to send more than 4 million boxes to landfills. Now, standard boxes can be used for 60 per cent to 80 per cent of all incoming parts, and boxes average eight uses. Box usage has been reduced by 2.4 to 3.2 million units a year; $1.5 million has been saved in pallet disposal costs. Payback was almost immediate.

Additional Waste Prevented

- Reduced freight cost because standard boxes and pallets "cube out" more efficiently.
- Reduced damage because boxes designed for reuse are sturdier; higher unit cost of sturdier boxes negated by large-volume purchases and reuse.
- Reduced storage costs owing to just-in-time delivery.

All things considered, Xerox estimates that the new packaging program saves its manufacturing facilities between $2 million and $5 million a year.

CASE STUDY 5 - INCOMING PACKAGING CONVERTED AS FILLERS

Corru-Fill / Corru-Shredder

2731 NE 14th Street

Pompano Beach, FL 33062

Summary

Incoming corrugated boxes (OCC) are converted into packaging filler for outgoing freight, saving the company $12,000 a year in polystyrene peanuts and OCC disposal costs—and creating a new profit center for the business.

Action

Norman Levine founded NDL Products, Inc., in 1974. The company (which Levine has since sold) manufactures and distributes products for the sports and fitness industries. Raw materials and components arrive from suppliers in corrugated boxes; finished products are shipped to customers in new boxes—some 250 to 400 cartons a day.

Before 1992, incoming corrugated boxes were disposed of, and polystyrene packing peanuts were used to fill voids in outbound boxes of products. "I was tired of all this corrugated going to the dumpster," Levine told an interviewer at the time. "I also wanted to stop using the polysty-rene peanut packing that was making a mess of our warehouse." The initial solution was to purchase a commercial cardboard shredder and chop the OCC boxes into pieces small enough to serve as packing material, replacing the polystyrene peanuts. But the pieces were too large, had a ragged look, and were full of small particles and paper dust. Customers complained. Levine and his associates then designed and built a new prototype which they called a Corru-Shredder, specifically to chop OCC pieces into uniform strips. The machine is 3 ft. wide, 8 ft. long, weighs about 400 pounds, and is mounted on wheels. It operates on 110 volts AC.

The operator prepares the OCC feedstock by flattening the cartons. Then, whole boxes are slit into strips, generally 4 to 6 inches wide, using a table saw. The width of this first cut determines the length of a finished piece of packing material. Next, the wide strips of OCC are fed into the Corru-Shredder for slicing into uniform, 1/8-inch-wide strips. The finished material, called Corru-Fill, passes through a strong air current to remove dust, and it falls into a bulk container. As the container fills, the contents are sprayed with a biodegradable insecticide in an alcohol-based carrier. Machine maintenance includes daily oiling and occasional sharpening of the four shear blades contained within. On an 8-hour shift with two operators, the machine can produce about 1,500 lbs. (1,000 cubic feet) of Corru-Fill, according to Levine.

By avoiding disposal of OCC and the purchase of polystyrene peanuts, the company recovered its cost to build the prototype Corru-Shredder in about 10 months. And since the machine was able to produce more Corru-Fill than needed for company shipping needs, there was a surplus to sell in the local market. Assuming a free supply of OCC, Corru-Fill is very competitive with polystyrene peanuts and is a superior packing material, Levine says, because it prevents the product from shifting during shipment.

Additional Waste Prevented

For equal volumes, chopped OCC is significantly heavier than foam polystyrene. However, when Corru-Fill is dumped loose around products in cartons, the exposed corrugations on the thin strips of OCC cause adjacent pieces to catch on one another, forming a sort of lattice, Levine says. The many air spaces thus created tend to negate weight differences between OCC and polystyrene, as well as reduce the required volume of packing material.

CASE STUDY 6 - PLASTIC PALLETS INSTEAD OF WOODEN PALLETS Pepsi-Cola Bottling of Phoenix

4242 East Raymond Street Phoenix, AZ 85040

Summary

In cooperation with one of its can suppliers, Pepsi now receives half of its deliveries of empty cans on plastic pallets instead of wooden pallets, eliminating 150 tons of wood pallet waste a year and reducing downtime and labor expense associated with wooden pallets.

Action

Since 1992, Pepsi-Cola Bottling of Phoenix has replaced half of its internal inventory of 44 x 56-inch wooden pallets with plastic pallets of the same dimension. Pepsi and the can supplier implemented the program after observing numerous problems associated with the delivery of can stock on wooden pallets. Slightly damaged or misshaped pallets sometimes splintered during automated unloading, interrupting production and causing substantial losses of new cans headed for the filling operation.

Pepsi asked its can suppliers to switch to plastic pallets, and one did. As a result, the plant manager reports: "Raw material losses resulting from pallet damage have fallen dramatically, labor to repair pallets for reuse has been reduced, and we have cut our production of wood waste by over 50 per cent." He said the conversion from wooden to plastic pallets was easy and did not require any operational changes.

Payback

The conversion from wood to plastic cost Pepsi nothing except discussions and planning with its supplier; therefore, payback to the bottler was virtually immediate. For the can suppliers, the investment in plastic pallets was recovered in less than 3 years.

Pepsi lists the following specific savings:

1. Labor cost to clean and repair damaged wooden pallets reduced by $3,000 a year
2. Wood waste generation cut in half
3. Production downtime (machine interruptions as a result of stoppage on the line stemming from defective pallets) cut from 10 hours a month to about half an hour
4. Useful life of pallets extended to at least 3 years or at least 30 trips, at least four time the performance of wooden pallets.

   It's important to note that Pepsi-Cola Bottling of Phoenix is using plastic pallets only within its own bottling plant. The company studied but decided not to use plastic pallets for shipment of product outside the plant, citing a number of reasons. According to the company, plastic pallets are:

4. Much more costly than wooden pallets
5. Hard enough to retain and control within the internal system let alone permitting them outside
6. Not yet in wide use outside and therefore something of a curiosity and collector's item, aggravating the problem of getting plastic pallets back.

CASE STUDY 7 - USING STANDARD REUSABLE CONTAINERS NACHI Technology, Inc.,

713 Pushville Road, P.O. Box 250 Greenwood, IN 46143

Summary

A supplier of ball bearings to the auto industry reduces pallet, box, and expanded polystyrene (EPS) purchase costs by $25,600 a year—and disposal costs by $6,800 a year— by working with raw materials suppliers and customers, including GM, to adopt standard and reusable logistical shipping materials.

Action

NACHI Technology, Inc., manufactures approximately 60 per cent of the air conditioning compressor bearings used in passenger and light trucks in the United States. The company employs 80 people and has sales of about $20 million a year. As a key supplier to the automotive industry, the company is acutely aware of the need to control costs. Recently it focused upon packaging as a possible way to reduce materials and disposal costs and improve profit margin.

Originally, customer GM required both a special pallet size (32-3/4 x 40-3/4") and a special corrugated box size (19-1/2 x 12-1/2 x 6") for packing bearings. After discussions with raw material suppliers and the customer, NACHI found that a pallet of more standard size—39 x 45"—and a "NACHI Box" measuring 14-3/4 x 9-1/2 x 6" would be acceptable for both delivery of raw material and shipment of finished bearings.

In addition, NACHI found that it could reuse EPS packaging in incoming freight as cushioning for outbound freight, eliminating the need to purchase 100,000 EPS packing blocks per year.

NACHI also sought and received approval of another customer to use an incoming corrugated carton containing rubber seals as the outgoing carton for finished bearings, reducing the need for new boxes by a further 720 boxes a month.

NACHI's principle savings can be summarized as follows:

- Savings on new pallets for shipments to GM $ 4,600
- Savings on new pallets for shipments to Ford 5,000
- Savings on corrugated boxes to GM 11,000
- Savings on polystyrene packaging blocks, GM 5,000
- Reduction of non-hazardous compactor waste 3,800
- Disposal savings from recycling old cartons 3,000
- Total annual savings $32,400 Payback

Payback must be considered virtually immediate since little investment was required outside of negotiating time with suppliers and customers. Two additional benefits should be noted:

- Corrugated boxes used to ship finished bearings originally contained 252 units and weighed about 100 pounds, well in excess of manual lifting limits. The new bulk pack is smaller and weighs about 40 pounds, facilitating the presentation of material at assembly points.
- NACHI donates serviceable used pallets to a nearby cooperative, avoiding disposal costs for the company and pallet purchase costs for the co-op.

CASE STUDY 8 - REPLACE SHRINK WRAP OF PALLETS BY RUBBER BANDS Traex/Menasha Corporation, Traex Plaza

P.O. Box 200, Dane, WI 53529-0200

Summary

Rubber pallet bands replace shrink wrap for securing skid-loads in warehouse operations, yielding annual savings of $49,706 from reduced labor and materials.

Action

Traex manufactures injection molded plastics and other products for the food service industry.

Finished products are boxed in corrugated containers and stored on pallets in the warehouse until needed to fill orders. Previously the company stretch-wrapped the pallet-loads by hand for transfer to the warehouse. The wrapping process was time-consuming and required most of one employee's attention over three shifts. The process also exposed workers to the risk of back injury from stooping and straining.

Traex began to study a change in this method of warehousing on pallets when its waste hauler announced a new requirement: old stretch-wrap removed from pallets during the order-filling process henceforth would have to be baled for disposal.

Beginning in April 1995, Traex began testing rubber pallet bands to secure loads. (Aero Pallet Bands, 7501 W. 99th Place, Bridgeview, IL 60455. 800-662-1009, 708-430-4900. Fax: 708-430-4909.) Unstretched bands measure 3/4" wide by 92" in circumference, the proper size for use with 40" x 48" pallets. Typically, two bands are used per skid, one looping around the top layer of boxes and the other about mid-load. Bands are applied by hand, an operation that takes one worker about 10 seconds per band. (By comparison, stretch-wrapping required about 1 minute per skid.) When skids are brought from the warehouse to fill orders, the bands are quickly removed and tossed in a container for reuse. From pallet-loading at the production line, to warehouse, to pallet-unloading, bands make the roundtrip in about six weeks.

With training provided by the warehouse manager, this change in operations was adopted easily. Employees like the rubber bands better than shrink-wrap. After 10 months under test, only four bands broke. There is no risk from snapping bands.

Payback

The most important saving is reduced labor cost. Traex calculates that it recovered the cost of its initial supply of rubber bands in just four 3-shift production days. Here is the company's summary of overall annual savings:

- Reduced labor $50,112
- Reduced disposal expense 400
- Reduced purchase of stretch-wrap 195
- 550 rubber bands @ 1.82 - 1,001 NET SAVING $49,706.

Index

Q

R

S

T

U

V

W